MITCHELL'S
Movement Control in the Fabric of Buildings

Philip Rainger *ARIBA*

Batsford Academic and Educational Ltd   London
NP Nichols Publishing Company   New York

© *Philip Rainger 1983*

All rights reserved. No part of this publication may be reproduced, in any form or by any means, without permission from the Publisher

First published in Great Britain 1983

Typeset by Keyspools Ltd, Golborne, Lancs
and printed in Great Britain by
R J Acford, Chichester, Sussex
for the publishers
Batsford Academic and Educational Ltd
an imprint of B T Batsford Ltd
4 Fitzhardinge Street
London W1H 0AH

ISBN 0 7134 3508 9

First published in USA by Nichols Publishing Company
P O Box 96, New York, NY 10024

Library of Congress Cataloging in Publication Data

Rainger, Philip.
   Movement control in the fabric of buildings.

   Includes bibliographical references.
   1. Structural engineering. 2. Buildings. I. Title.
TH845.R34   1983   690′.21   83–8283
ISBN 0–89397–168–5

**Movement Control in the
Fabric of Buildings**

Contents

ACKNOWLEDGMENT 8

PREFACE 9

SECTION 1 INTRODUCTION 10

SECTION 2 SOURCES AND EFFECTS OF MOVEMENTS: GENERAL PRINCIPLES 11

2.1 Identification of sources 11

2.2 Classification of sources 11

2.3 Principal causes of movements and effects 11

2.4 Effects: deformations and stresses 11

2.5 Estimation of movements 15
Loading effects: elastic deformations due to loading 15
Thermal movements 17
Moisture movements 18

2.6 Estimation of stress due to restraint 18

SECTION 3 MOVEMENT CHARACTERISTICS OF MATERIALS 23

3.1 Scope 23

3.2 Basic data 23

3.3 Concrete 23
Early age movements in concrete 23
Deformation movements in the hardened concrete 30
Concrete blockwork 34

3.4 Bricks and brickwork 34

3.5 Timber 36

3.6 Ceramics 41

3.7 Metals 41

3.8 Plastics 42

SECTION 4 DESIGN FOR MOVEMENT IN BUILDINGS 43

4.1 Relative movement problems 43
Alternative strategies 43
Related design problems 44
Assessment of movements and inaccuracies 44

4.2 Design for movement: summary of procedure 44

4.3 Movement accommodation 44

4.4 Movement joint design 46
Procedure in joint design 46
Calculation of movement joint widths 47
Joint design and details 57

SECTION 5 DESIGN FOR MOVEMENT CONTROL IN BUILDING ELEMENTS 66

- 5.1 **Foundations** 66
  - Scope 66
  - Design criteria and failure limits 66
  - Sources of movement: soil and superstructure 67
  - Summary of causes of non-load related movements 71
  - Movement control methods 71
  - Selection of foundation types (in shrinkable clay soils) 75

- 5.2 **Basements and substructures** 76
  - Scope 76
  - Ground floor slabs: design criteria 79
  - Ground floor slabs: causes of movement 79
  - Concrete ground floor slabs: causes of movement 83
  - Small floor slabs ($\not> 7m \times 7m$): movement control 83
  - Design and construction methods for movement control in large area non-structural concrete slabs on the ground 86
  - Retaining walls and free-standing walls 90
  - Basements: design criteria 90
  - Movement failure Limits 90
  - Basement construction: movement control methods 90

- 5.3 **External works** 95
  - External paved areas: design criteria 95
  - External free-standing walls 97

- 5.4 **Structural frameworks** 97
  - Scope 97
  - Design criteria and failure limits 97
  - Deformation limits 100
  - Structural framework: causes of movement and effects 103
  - Movement control methods 103
  - Accommodation of movement 105
  - Movement joints: concrete structures 106

- 5.5 **Walls: brick and block masonry** 108
  - Scope 108
  - Movement control criteria 108
  - Assessment of movement in walls 112
  - Location: restraint 113
  - Deformation in adjoining elements 118
  - Surface characteristics: severity of exposure 118
  - Location of critical points in walling 122
  - Identification of causes of movement and prediction of effects 122
  - Estimation of effects 124
  - Selection of movement control methods 125
  - Movement control methods in walls: alternative (1) accommodation 125
  - Selection of wall jointing materials 133
  - Provision of wall movement joints 133
  - Summary of wall joint types 138
  - Movement control methods in walls: alternative (2) mitigation/reduction and effects 140

- 5.6 **External wall claddings and facings** 145
  - Movement in claddings and facings 145
  - Scope 145
  - Movement control criteria and failure limits 147
  - Movement characteristics of facing materials 147

Assessment of movements and
  effects   147
Location and direction of
  deformations due to movements in
  the supporting structure   150
Assessment of movements and
  deformation in claddings, facings   152
Detailed design to allow for
  movement   152

5.7 Roofs and roof finishes   159
  Scope   159
  Design criteria   159
  Failure limits   160
  Causes and effects of movements in
    roofs   160
  Movement control methods and
    procedure   162
  Movement in sub-strate/roof-decks   164
  Movement control methods in roofs   167

5.8 Internal suspended floors   169
  Scope   169
  Classification of floor types   169
  Design criteria for floors   169
  Causes of movements/deformations
    in floors   175
  Timber floors   183
  Concrete floors   184
  Floor movement joints: design
    procedure   184
  Floor joint sealants and fillers: main
    types   187

5.9 Internal finishes   190
  Floor finishes   190
  Design criteria movement failure
    limits   190
  Causes of movements and
  effects   191
  Failures in flexible floor finishes   191
  Movement control methods in
    floors   191
  Internal wall finishes   194
  Scope   194
  Design criteria   196
  Causes of movement   196
  Movement control methods:
    wet plaster finishes   196
  Wet bedded wall tiling   196
  Framing for dry fixed linings/
    plasterboard   197

5.10 Subsidiary components   203
  Scope   203
  Design criteria   203
  Glazing systems   203
  External joinery: draught
    exclusion   204
  Mechanical services   204

SECTION 6   GENERAL
REFERENCES TO ALL
SECTIONS   210

RELEVANT NATIONAL STANDARDS

  American National   213
  Federal Republic of Germany   213

Index   214

Acknowledgment

My thanks are due primarily to the authors of the excellent technical guide books referred to in the Reference sections for concise, relevant information in the various fields covered by this volume, principally the authors of the publications by the following Associations, from whom copies of the literature referred to in the text and bibliography can be obtained:

Brick Development Association (BDA) Woodside House, Winkfield, Berkshire SL4 2 SX;

Cement and Concrete Association (C&CA), Wrexham Springs, Slough SL3 6P1;

Structural Clay Products Ltd; Peck House, 95 Fore Street, Hertford, Herts SG14 1AS.

I am grateful for the personal help and guidance given by Robert Wilson of the Cement and Concrete Association, and the Learning Resources staff at Brighton Polytechnic for finding and obtaining much of the reference material.

I am also grateful for the kind permission of the Directorate of Architectural Services of the Property Services Association (PSA) for permission to use diagrams from their *Flat Roofs* Technical Guide, available from the PSA Library Sales Room C109, Whitgift Centre, Croydon, Surrey.

My thanks too for the many useful sources in technical guidance given by sealant specialists and numerous trade associations.

Finally, my thanks to Franklyn Nevard, BA(Hons), Dip Arch, whose work on the illustrations has been indispensible.

Brighton 1983   PR

Extracts from British Standards are reproduced by permission of the British Standards Institution (BSI), 2 Park Street, London W1A 2BS. Complete copies of the Standards can be obtained from the BSI Sales Department, 101 Pentonville Road, London N1 9ND.

Building Research Establishment (BRE) (including Princes Risborough Laboratory) publications, from which information is extensively used, are identified throughout the text. This material is contributed by courtesy of the Director, BRE and is reproduced by permission of the Controller, HMSO. Crown copyright.

Preface

The problem of failures in buildings and the increase in claims against architects and designers are causing increasing concern both to the public and to the professions. A great deal of excellent authoritative guidance has been published, based on solid research and experience of defects found in the field.

The problem for the designer is not so much a shortage of information but the sheer volume of the guidance available. The information also tends to be specialised, ie dealing with limited areas only.

The Building Research Establishment in its Digest 176 *Failure patterns and implications*, states that the 'major shortcoming in design appeared to be failure to make use of authoritative design guidance'. It attributes this both to the sheer volume of guidance available and to the designer's desire to work out each solution for himself as a new problem.

The aim of this book is to focus attention on *movement*, which is one of the main sources of failure in buildings, and to provide architects and designers with guidance notes to cover all aspects of the problem in the fabric of buildings. Other sources of failure, such as water penetration and vapour control, require separate attention.

Structural engineers have been dealing with movements in structures, from large bridge spans to smaller buildings, for a considerable time and the Codes of Practice show the result of a great deal of research and experience in this field.

However architects have not always appreciated the importance of a close integration between the movements and movement joint requirements of the structural framework and the rest of the fabric. It is hoped that by explaining some of the movement problems of concrete in this book architects and designers will be able to better appreciate these problems and see the importance of a closer working relationship at design stage with all concerned.

Although the initial choice of building fabric is made from a broad based assessment of requirements and limitations, nevertheless at quite an early stage in the choice and detailed design of assembly techniques it is necessary to take account of the need to control the effects of movements. It is hoped that this book will be not only a handy reference during design but also a useful source of background information when approaching the problems of the choice and detailing of the fabric. It is also hoped that it will prove useful to students in an appreciation of the principles of building fabric design.

1 Introduction

Trends in present day construction methods have caused an increase in failures due to movements. This is due to the following factors:

1 The use of thinner sections with a lower thermal capacity, resulting in higher service temperatures in the structure.
2 Larger units, with greater movement at fewer joints.
3 The use of new materials, eg plastics with larger movement potential.
4 More highly stressed, slender structural members and the use of prestressing.
5 The increasing use of dry jointed assemblies.

It is now generally recognised that consideration at the design stage and anticipation of the potential movements in the various materials and components of the building fabric is essential in order to prevent failures in the body of the materials themselves, in the joints and the fixings.

The long term performance of buildings and the need for maintenance must be considered in the design of movement control methods.

The various causes of movement are often interactive and difficult if not impossible to predict precisely. They must also form part of the analysis of all stresses in the building, so that collaboration between all concerned in building design is essential for the successful integration of the provisions for movement in the fabric.

There is a need to evolve a strategy for movement control in each individual case to suit the exposure, form and materials of the building. If this is not done and successfully carried out the building will find its own movement accommodation but not always where desired by the designer.

The purpose of this book is to provide a guide for the identification and prediction of movements at design stage and to give information on detailing and the execution of movement control methods.

2 Sources and effects of movements: general principles

2.1 IDENTIFICATION OF SOURCES

The estimation of likely movements and related stresses is part of the design procedure described in section 4.

Before any assessment can be made of the location and extent of movements in the building, the possible causes of movements must be identified. It is helpful to be aware of some general classifications of sources as follows:

2.2 CLASSIFICATION OF SOURCES OF MOVEMENTS

Extrinsic These can be external (extrinsic), ie due to the factors imposed on the building, eg loads, solar heat gain changes in external temperature humidity, or
Intrinsic internal (intrinsic) causes due to changes occurring in the materials of the fabric, eg moisture content changes chemical and physical changes in the materials.
Time related Seasonal: (extremes of summer/winter)
diurnal: (twice daily)
exceptional:
instant: eg on loading
initial or recurring.

Table 2.1 gives the basic causes classified as described.

2.3 PRINCIPAL CAUSES OF MOVEMENTS AND EFFECTS

See table 2.1 page 12.

2.4 EFFECTS: DEFORMATIONS AND STRESSES

Induced stresses and deformation
If volume changes, due to whatever cause, cannot take place freely, then this restraint causes internal stresses and strains leading to deformations other than those due to simple volumetric changes.

So the same causes can result in stress and deformation if they are restrained or simple movement if they are not.

Degree of restraint

The amount of movement actually taking place will also depend on the degree to which this movement is restrained. The relationship can be summarised as follows:

$$\text{Cause of movement} \rightarrow \begin{array}{c}\text{Free movement}\\+\\\text{Restraining force}\end{array} = \begin{array}{c}\text{Restrained movement}\\+\\\text{Stress}\end{array}$$

Restraining force

The force required to restrain the movement is obviously dependant on the amount of free movement to be restrained and the modulus of elasticity of the material.

The restraining force provided by say a fixing, needs to be known for the correct design to avoid failure.

CAUSES OF STRESS

Stresses are caused whenever a material is restrained from taking up the 'potential' change of volume/size or shape. This can be **external** by fixings/supports or adjacent materials or due to **internal** causes, which usually occurs in laminated, composite or non-uniform materials. In some cases both forms of stress may occur simultaneously.

Types of stress

This will depend on the *potential deformed* shape

SOURCES AND EFFECTS OF MOVEMENTS: GENERAL PRINCIPLES

Table 2.1

Type	Cause	Time and duration	Unrestrained volumetric changes
Extrinsic	External climatic temperature changes Solar radiation	Exceptional Intermittent Seasonal Diurnal	— Expansion Contraction
	Ambient humidity changes Wetting Drying Moisture content changes	Seasonal Exceptional Short term Initial Alternate wetting/drying	Shrinkage Expansion
Intrinsic	Loading Dead loads	At time of erection Permanent	Immediate: elastic deformations Progressive: creep
	Live loading (applied)	Permanent or Intermittent	Elastic and non-elastic deformations
	Wind loading	Intermittent	Deformations
	Vibrations	Recurring	Oscillations
	Chemical changes Loss of volatiles	Initial	Contraction
	Corrosion and sulphate formation	Continuous Progressive	Expansion
	Physical changes ice formation loss of moisture	Intermittent Recurrent	Expansion Shrinkage

which has been restrained. Restraint of simple linear changes, as shown in diagram 2.1 result in tension and compression, whereas materials that would have been subject to curvature will have bending and shear stresses induced in them.

Magnitude of stress

The magnitude of the stresses will be proportional to the difference in the strain (change of size) between the deformed or undeformed shape and these stresses can be computed.

Acceptable stress limits

BRE Digest 227 makes the following general comment on the acceptability of various stresses:

'To ascertain whether a given level of stress is acceptable, it will also be necessary to know the yield stresses or permissible stresses of the material in the modes of stress to which it is subject: compression, tension, bending, shear, etc. This information is usually specific to a particular grade or variety of material. As the strength of many building materials, such as masonry and concrete

EFFECTS: DEFORMATIONS AND STRESSES

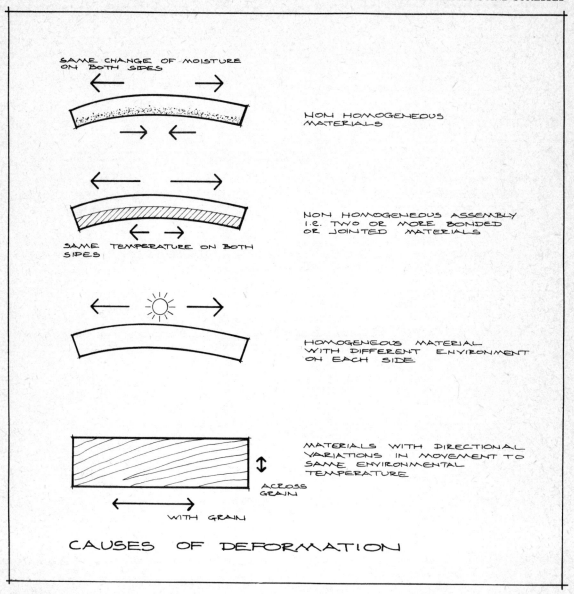

Diagram 2.1

is far less in tension than in compression, quite low stresses may give rise to cracking where there is restraint to contraction. The acceptability of this will vary in different situations, according to the criteria governing appearance, durability and other functions.'

Direction of deformation

Moisture and thermal changes are causes of movement, described earlier, causing simple changes of volume if completely unrestrained, acting on a solid in all principal directions. Applied loading causes a resistance directly opposed to the direction in which it is applied and corresponding shortening and lengthening in compression or tension, but there is an associated change of size in the opposite direction often at right angles. See diagram 2.2.

Deformations other than simple volumetric

13

SOURCES AND EFFECTS OF MOVEMENTS: GENERAL PRINCIPLES

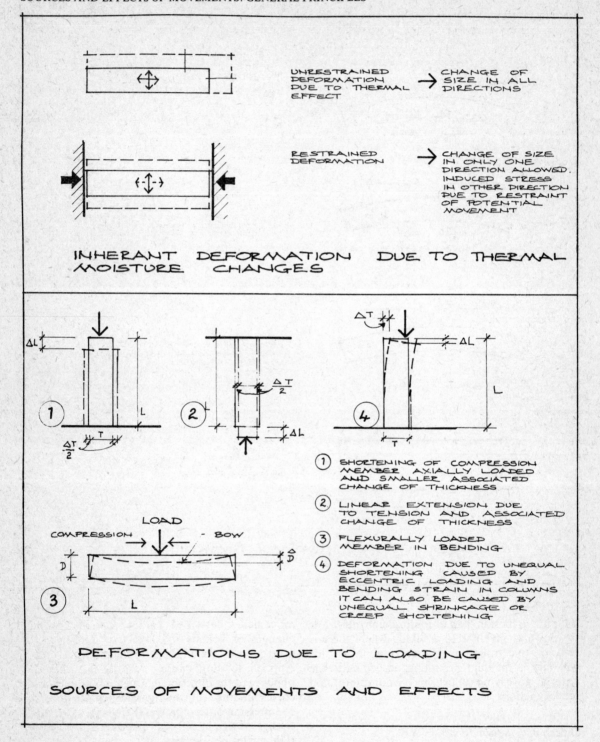

Diagram 2.2

changes also occur for the following reasons. These are also illustrated in diagram 2.1.

1 Non homogeneous materials or assemblies with different movement characteristics.
2 Variations in environmental conditions across the material or component or on each side.

In both cases the deformations are restrained within the material or assembly, thus causing 'locked-in' stresses.

2.5 ESTIMATION OF MOVEMENTS

Free and unrestrained movements/deformation evaluation

After identifying the location and direction of all potential movements, it is necessary to recognise all possible restraints so that any deformations and stresses can be predicted as well as the free movements if any.

As a first step the *free* movement of the material of construction or component can be evaluated, ie the unrestrained movement. This may be adjusted by approximations to arrive at the actual final movement or more precise methods of calculating stresses may be required.

Simple 'free to move' components

In the case of simple components more or less free to move between flexible joints, eg cladding panels, the free movement calculation may be sufficient for joint design. However, care must be taken to predict deformations as for example bowing due to variable material composition (backed panels) on environmental conditions.

Combined movements: effects

Most building components are subject to more than one cause of movement, the effects of which need to be added to obtain the combined total movement.

This is essential before the final design steps as described in section 4, namely adding the allowances for tolerances jointing to arrive at final joint design widths and a strategy for movement control.

This section describes the techniques of calculating movements. It must be read with the movement data for materials in section 3, and applied in the design procedure described in section 4.

LOADING EFFECTS: ELASTIC DEFORMATIONS DUE TO LOADING

These deformations form part of the Structural Engineering design of the building and are not normally carried out by the fabric designers. However, it is useful to appreciate that the elastic properties of materials vary greatly and that other effects are caused by loading in addition to elastic deformations.

Also at various levels of stress, materials are no longer truly elastic.

A truly elastic deformation is one which obeys Hooke's law that stress is proportional to strain and this is a constant ratio for whatever level of stress, ie

The modulus of elasticity (Young's modules) is this constant of $\frac{\text{stress}}{\text{strain}}$ for any individual material.

It applies only where the material is truly elastic.

Modulus of elasticity and non-elastic materials

Many important building materials are not truly elastic and their characteristics are as described in section 3 under each material and as illustrated, in diagram 2.3.

Elastic deformations: calculations

Some typical elastic deformations are shown in diagram 2.2. The amount of deflection can be directly derived from the area of the bending movement diagram, the elastic modulus and moment of inertia of the section. (For full statement see D. Lenzner: *Movement in Buildings* and textbooks on structural mechanics.)

Load related non-elastic deformations

These are closely related to the properties of non elastic materials and described under the appropriate part of section 3.

SOURCES AND EFFECTS OF MOVEMENTS: GENERAL PRINCIPLES

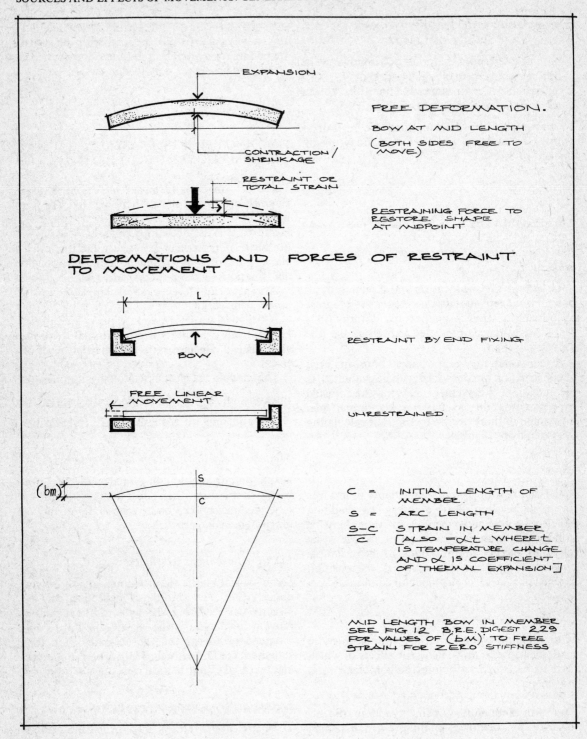

Diagram 2.3

ESTIMATION OF MOVEMENTS

THERMAL MOVEMENTS

Co-efficient of thermal expansion

Change of temperature in a solid, causes a change of size. The rate is a constant for each material, or the amount of this change per degree change of temperature for any material is a constant. It is expressed as a *proportion* of *change* to *original size*. This is called the co-efficient of thermal expansion.

To obtain the change of size in any component simply multiply the original size in *any* unit of length by this co-efficient and the degree of temperature change, ie

$R = \alpha L t$
R = change of size
α = co-efficient of thermal expansion (per 1°C)
L = length of section considered
t = temperature *difference*, ie *number of degrees of change*.

The table of co-efficients of linear thermal expansion are given in section 3.

Assessment of temperature changes

The *air* temperature in Great Britain can vary from $-5°C$ to $30°C$ (35 K) but the actual temperature of materials subjected to radiation varies greatly and depends on colour and degree of protection. Table 2.2 taken from BRE Digest 228 gives the range of service temperature to be taken into account.

Table 2.2 Examples of service temperature ranges of materials (valid for UK only)

	Min °C	Max °C	Range °C
External			
Cladding, walling, roofing			
Heavyweight			
Light colour	−20	50	70
Dark colour	−20	65	85
Lightweight, over insulation			
Light colour	−25	60	85
Dark colour	−25	80	105
Glass			
Coloured or solar control	−25	90	115
Clear	−25	40	65
Freestanding structures or fully exposed structural members			
Concrete			
Light colour	−20	45	65
Dark colour	−20	60	80
Metal			
Light colour	−25	50	75
Dark colour	−25	65	90
Internal			
Normal use	10	30	20
Empty/out of use	−5	35	40

The following situations are not included in the above examples, and may give rise to temperature extremes more severe than those listed:
Dark surfaces under glass, eg solar collectors.
Materials used in cold rooms or refrigerated stores.
Materials used for, or in proximity to, heating, cooking and washing appliances, or flues and heat distribution networks.

Base temperatures for thermal change calculations

The temperature at time of construction will give the basis for the allowance to be made for subsequent changes. As most buildings are assumed to be built at an average of 10°C, so, for example in claddings, the amount of the *contraction* $+10°$ to $-20°C$, ie a 30° change, is probably *less* than the *expansion*, ie $+10°C$ to $+50°C$, ie 40°C change so most movement joints are probably (expansion) joints.

Summary

The final thermal movement in practice will depend on:

1 Temperature range
2 Degree of exposure
3 Response of material (co-efficient E)
4 Colour of surface
5 Restraint to movement.

Differential temperature movements

It is essential to take account of different temperature exposures (eg aspects) in buildings and between different layers to assess the extent of the differential thermal movement effects.

Predicting final movement and restraint

This is more complex than calculating the free movement. Reference should be made to BRE Digest 228 for a fuller treatment of this subject which is given in principle in Section 2.6.

SOURCES AND EFFECTS OF MOVEMENTS: GENERAL PRINCIPLES

MOISTURE MOVEMENTS

Origins and types of moisture movement

These movements usually take place due to changes of size resulting from changes of moisture content in the material.

These changes depend both on the physical characteristics of the material, ie capability and rate of absorption or releasing of moisture and ambient conditions.

Initial permanent and irreversible changes

There are two types of moisture movements. Irreversible and reversible. The irreversible movements are usually due to the initial evaporation of mixing water or in the case of cement based mortars, renderings or concrete of the excess moisture in excess of the water required for hydration. The amount of water present and rate and quantity evaporating depends on mix design and ambient humidity and the shrinkage resulting is difficult to assess precisely. See section 3 on concrete for further details.

Reversible recurrent changes

Alternate wetting and drying *subsequent* to the initial moisture changes occur recurrently and are easier to assess.

BRE Digest 228 gives simple percentage factors (amount of movement expressed as a percentage of the original size).

These are based on the extreme moisture conditions or contents likely to occur in practice, ie normal building use.

The moisture contents are based on saturation (ie extreme wet) and extreme dry conditions, but in wood on moisture content equilibrium at stated percentages of moisture content.

Combined moisture and thermal effects

These are often not additive (eg hot dry weather). It is recommended therefore, that the dominant, ie greatest movement is calculated

    eg   concrete – shrinkage
           wood   – moisture movement

and other effects added or subtracted as appropriate.

2.6 ESTIMATION OF STRESS DUE TO RESTRAINT
(examples from BRE digests 227 and 228)

Induced stresses

In order to control movements to acceptable limits it is sometimes necessary to calculate the stress resulting from restraint to ensure that this falls within the strength of the material.

Restraining force

As the level of stress induced by restraint is independent of the cross sectional area of the member, the greater the area, the greater will be the force needed for restraint.

Force and stress due to axial restraint

Full axial restraint to deformation causes stress as follows:

$$\text{Stress} = E \times \text{strain}$$
$$\text{Force} = \text{Stress} \times \text{area}$$

For thermal movements,

the strain = α $\times$ t
          (co-efficient of   (temperature
           linear expansion) change)

and stress = $\alpha t \times E$
    force = $\alpha t E A$

Temperature at time of restraint

The restraint is often provided by a fixing or ends built in at the middle of the temperature range. So for example an aluminium mullion restrained at 10°C will expand up to 50°C and contract down to −20°C. Stress induced at *lower* extremes (30°C) would be

    stress = $\alpha \times t \times E$
           = 24 × 30 × 70
           = 50. 400. or <u>50 N/mm²</u>

and at upper limit (+10°C/+50°C = +40°C)

    stress = 24 × 40 × 70
           = 67,200 or <u>67 N/mm²</u>

Note The maximum permissable stress is 130 to 140 N/mm² so that buckling failure due to these additional stresses could easily occur.

ESTIMATION OF STRESS DUE TO RESTRAINT

Moisture movement restraint

For calculation of restrained moisture movements values of strain are equivalent to the percentage factors given in the table from BRE Digest 228, columns 3 and 4. Again they need to be divided by the conditions at which the restraint was imposed.

Induced stresses due to movement restraint

Table 2.3 gives examples of stresses induced by thermal movements (see also *Mitchell's Materials* by Alan Everett). The stress is the result of total restraint of the free movement strain caused by the temperature rises shown.

It should be noted that the stress induced in the brickwork in weak mortar is 3.9% of the failing stress, but the stress induced in the stronger brickwork is 6.6% of the failing stress.

Deformations and restraints other than linear

So far the methods shown are for *simple linear changes of size*. However, more complex deformations and restraining forces are common. These are shown in diagram 2.2 (a). For a detailed treatment of the calculations for the restraint reference should be made to BRE Digest 229 *The estimation of thermal and moisture movements: part 3*, from which the following is an example.

Bowing of restrained members: simple method of finding curvature

If a member is restrained by end fixings unable to accommodate the linear change of size, ie the strain, but free to rotate at the ends and is thin and flexible, bowing can take place to relieve the expansion stresses (see diagram 2.3). The chord length c is the original length and the arc length s the unrestrained freely expanded length, b m the amount of bow.

The strain is $\frac{s-c}{c}$ ($= \alpha t$ with t appropriately determined) without lateral resistant and for zero stiffness.

The amount of bow can then be derived from the graph given in BRE Digest 229 and reproduced on page 20.

Graphical design aids

In order to save time in converting movement strains into actual change of size of members, the following graph, published as technical guidance by the Cement and Concrete Association, should prove useful.

Additional effects due to conditions during erection

The effect of the dead and live loads can also vary

Table 2.3

Material	Modules of elasticity	Co-efficient of thermal expansion	Failing stress in compression	Stress due to 15 deg C rise in temperature
	N/mm²		N/mm²	N/mm²
Medium strength brickwork in cement mortar	6200	$6 c 10^{-6}$	180,000	12,000
Medium strength brickwork in weak (lime) mortar	1400	6×10^{-6}	71,000	2,800

SOURCES AND EFFECTS OF MOVEMENTS: GENERAL PRINCIPLES

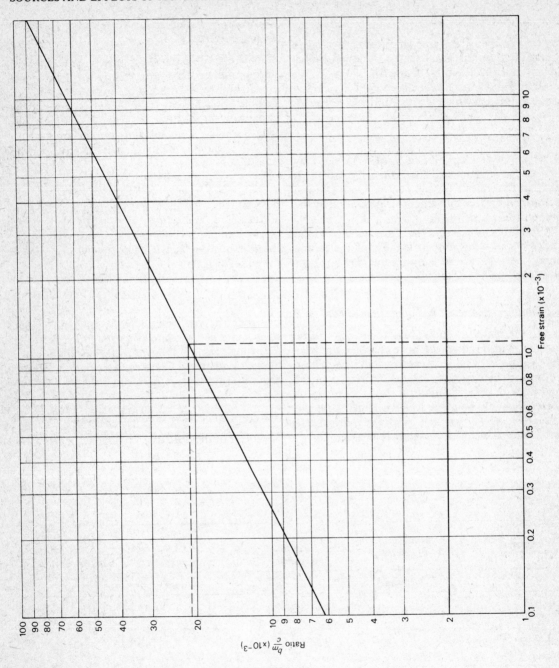

Relationship of bow to arc length and chord length
From BRE Digest 229

ESTIMATION OF STRESS DUE TO RESTRAINT

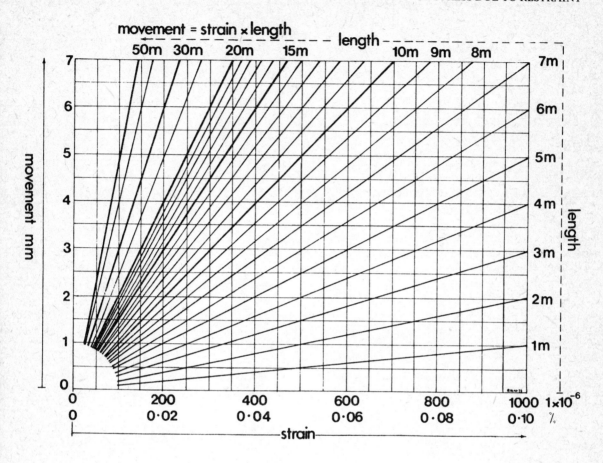

according to the following additive factors. These can combine either to reduce the total effect or to increase them with increased risk of failure:

Time of application of loads
eg partition can be build *before* or *after* additional dead loads causing initial deflections have been applied, eg floor screeds or when other movements have taken place, eg creep and initial shrinkage.

Temperature at erection
This is assumed to be +10°C but high or low temperature at construction can cause either shrinkage or expansion of the finished structure on completion *in addition* to normal temperature variations from inside/outside.

Atmospheric humidity condition during erection
This can cause movements either due to an increase or decrease in humidity. This is often differential due to variable conditions between interior and exterior, eg a column in the facade, external cladding panels, etc.

3 Movement characteristics of materials

3.1 SCOPE

This section gives the basic data on the movement characteristics of materials for use in the movement calculations described in section 2 and the jointing design procedures in section 4.

It is also intended as a comparative survey of properties to draw attention to differential movement characteristics of individual materials which have to be considered in design with a note of special precautions to be taken. Different causes of movement are critical for various materials and these must form the basis for the effects to be considered in each situation.

3.2 BASIC DATA

Differentials in movement potential

Tables 3.1 and 3.2 give the comparative and basic data for movements in materials.

Comparative data

Table 3.1 Compares the thermal movement characteristics of common materials.
Table 3.2 Compares the initial (irreversible) and recurrent (reversible) moisture movements of common building materials.

Basic data for movement calculations

Table 3.3 Gives the basic data on the movement characteristics of materials. This is reproduced from BRE Digest 228, part 2.

The following brief outline of the characteristics of materials is intended as a guide to select the critical movements likely to be encountered when these materials are used and to point to precautions to be taken to prevent failures.

Calculations are required for example in reinforced concrete where the final movement strains are a combination of many factors, including environmental conditions and time of erection and are part of the complete analysis of stresses in the structure.

3.3 CONCRETE

This material is well known as having very variable properties. These depend on mix design and composition, type of aggregates and water/cement ratio. Concrete movement is also dependant on the usual external factors of thermal and moisture change. The amount of non-elastic 'time related' movement (creep) depends also on both concrete composition and the time of application of loading. It is important first to draw the distinction between movement problems of the concrete before it is hardened and secondly of the hardened concrete and lastly, the various properties of different types of concrete. A useful summary of the movement characteristics of concrete is given in the C and CA publication concrete practice and the CIRIA technical note 107 given in the references.

EARLY AGE MOVEMENTS IN CONCRETE

See diagrams 3.1, 3.2 and 3.3

Heat of hydration movement

The setting action of concrete evolves heat. The concrete sets in the expanded condition and on cooling shrinks. This shrinkage occurs first in the outer layers and is restrained by the interiors and the resulting stress causes the cracks. Engineers are now able to predict this type of cracking and take precautions.

MOVEMENT CHARACTERISTICS OF MATERIALS: COMPARATIVE TABLES

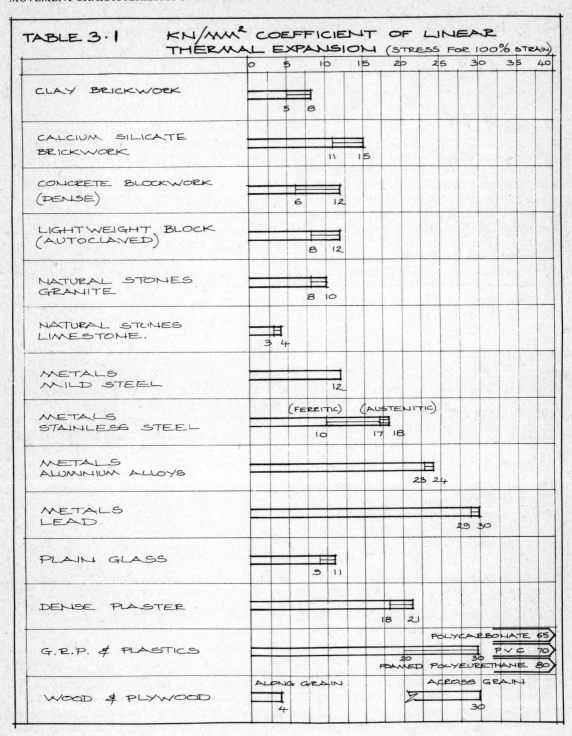

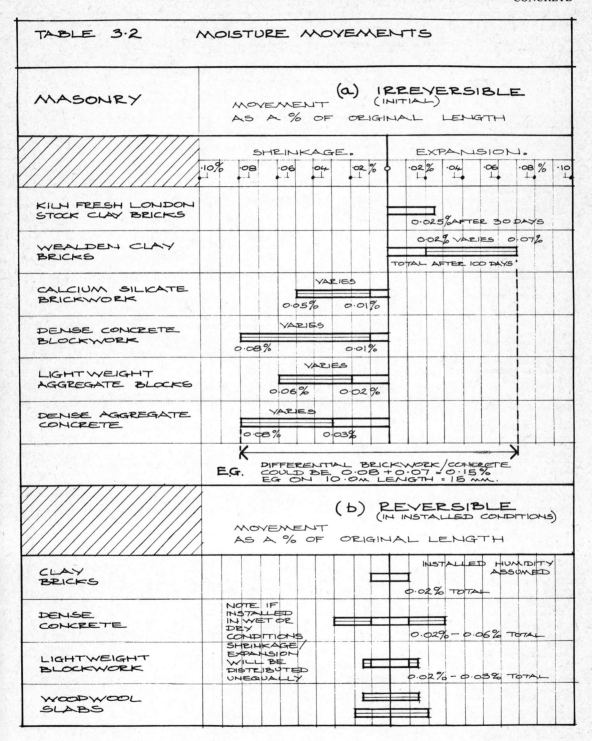

MOVEMENT CHARACTERISTICS OF MATERIALS

Table 3.3 Properties needed to assess changes of size and shape of materials

Note: Unless more specific data are available, design should be based on the higher value where a range is shown

(1) Material	(2) Coefficient of Linear thermal expansion α per $°C \times 10^{-6}$	(3) Reversible moisture movement %	(4) Irreversible moisture movement (+) expansion (−) shrinkage %	(5) Modulus of elasticity E kN/mm^2
Natural stones				
Granite	8–10			20–60
Limestone	3–4	0.01		10–80
Marble	4–6			35
Sandstone	7–12	0.07		3–80
Slate	9–11			10–35
Cement-based composites				
Mortar and fine concrete	10–13	0.02–0.06	0.04–0.10 (−)	20–35
Dense aggregate concrete				
Gravel aggregate	12–14	0.02–0.06	0.03–0.08 (−)	15–36
Crushed rock (except limestone)	10–13	0.03–0.10	0.03–0.08 (−)	15–36
Limestone	7–8	0.02–0.03	0.03–0.04 (−)	20–36
Steel-fibre reinforced concrete	5–14	0.02–0.06	0.03–0.06 (−)	20–41
Aerated concrete	8	0.02–0.03	0.07–0.09 (−)	1.4–3.2
Lightweight aggregate concrete:				
Medium lightweight	8–12	0.03–0.06	0.03–0.09 (−)	8
Ultra-lightweight (expanded vermiculite and perlite)	6–8	0.10–0.20	0.20–0.40 (−) (values for plain concrete; moisture movements may be partially restrained by appropriately placed reinforcement)	†
Asbestos-cement	8–12	0.10–0.25	0.08 (−)	14–26
Glass reinforced cement	7–12	0.15–0.25	0.07 (−)	20–34
Calcium silicate based composites				
Asbestos wallboard and substitutes	5–12	0.14–0.27		8–10
Asbestos insulating board and substitutes	2.5–7.2	0.16–0.25		2.6–3.6
Gypsum and gypsum-based composites				
Dense plasters; plasterboard	18–21			16
Sanded plasters	12–15			8.5–16
Lightweight plasters	16–18			1.5–4
Glass-reinforced gypsum	17–20			16–20
Brickwork, blockwork and tiling				
Concrete brickwork and blockwork:				
Dense aggregate	6–12	0.02–0.04	0.02–0.06 (−)	10–25
Lightweight aggregate (autoclaved)	8–12	0.03–0.06	0.02–0.06 (−)	4–16
Aerated (autoclaved)	8	0.02–0.03	0.05–0.09 (−)	3–8
Calcium silicate brickwork	8–14	0.01–0.05	0.01–0.04 (−)	14–18
Clay or shale brickwork or blockwork	5–8	0.02	0.002–0.01 (+)	4–26
Clay tiling	4–6	†	†	†
Metals				
Cast iron	10			80–120
Plain carbon steel	12			210
Stainless steel:				
Austenitic	18			200
Ferritic	10			200
Aluminium and alloys	24			70
Copper	17			95–130
Bronze	20			100
Aluminium bronze	18			120
Brass	21			100
Zinc	33 parelle to rolling / 23 perpendicular to rolling			140 parallel to rolling / 220 perpendicular to rolling
Lead	30			14
Wood and wood laminates*				
Softwoods	4–6 with grain / 30–70 across grain	0.6–2.6 tangential[1] / 0.45–2.0 radial[1]		5.5–12.5[4]

continued

CONCRETE

continued

(1) Material	(2) Coefficient of Linear thermal expansion α per °C × 10^{-6}	(3) Reversible moisture movement %	(4) Irreversible moisture movement (+) expansion (−) shrinkage %	(5) Modules of elasticity E kN/mm^2
Hardwoods	4–6 with grain 30–70 across grain	0.08–4.0 tangential[1,5] 0.5–2.5 radial[1]		7–21[4]
Plywood	†	0.15–0.20 with grain[2] 0.20–0.30 across grain[2]		6–12[4]
Blockboard and laminboard	†	0.05–0.07 with core[2] 0.15–0.35 across core[2]		7–11 with core[4] 5–8 across core[4]
Wood-chip and fibrous materials		(On length or width; values for thickness may be up to 30 times greater)		
Hardboard	†	0.30–0.35[2]		3.0–6.0
Medium board	†	0.30–0.40[2]		1.7–3.3
Softboard	†	0.40[2]		†
Chipboard	†	0.35[3]		2.0–2.8
Wood-wool-cement	†	0.15–0.30 on length 0.25–0.40 on width	†	0.6–0.7
Rubbers and plastics etc				
Asphalt	30–80			†
Pitch fibre	40	0.2–0.3		†
Ebonite	65–80			†
Thermoplastics:			While not subject to moisture effects, some plastics may be liable to irreversible progressive shrinkage due to loss of volatiles and related causes.	
PVC UPVC CPVC	40–70			2.1–3.5
Polyethylene (low density)	160–200			0.1–0.25
(high density)	110–140			0.5–1.0
Polypropylene	80–110			0.9–1.6
Polycarbonate	60–70			2.2–2.5
Polystyrene	60–80			1.7–3.1
Acrylic	50–90			2.5–3.3
Acetal	80			2.8–3.7
Polyamide	80–130			1.0–2.7
ABS	60–100			0.9–2.8
Thermosets (laminates):				
Phenol and melamine formaldehyde	30–45			5.5–8.5
Urea formaldehyde	27			10
Cellular (expanded):				
PVC	35–50			†
Phenolic	20–40			†
Urea-formaldehyde	30–40			†
Polyurethane	20–70			†
Polystyrene	15–45			†
Reinforced:				
GRP (chopped strand)	20–35			6–12
Carbon-fibre (orientated)	0–0.05 parallel to reinforcement 30–70 perpendicular to reinforcement			180–220
Glass				
Plain, tinted and opaque	9–11			70
Foamed (cellular)	8.5			5–8

[1] Based on 60% and 90% relative humidities.
[2] Based on 33% and 90% relative humidities.
[3] Based on 65% and 90% relative humidities.
[4] at 12% moisture content; values reduced at higher moisture contents.
[5] Negligible with grain. Noted movements are across grain.
* More specific data can be found in PRL Technical Note 38, *The movement of timbers* and PRL Bulletin No. 50, *The strength properties of timbers*.
† No date available.

MOVEMENT CHARACTERISTICS OF MATERIALS

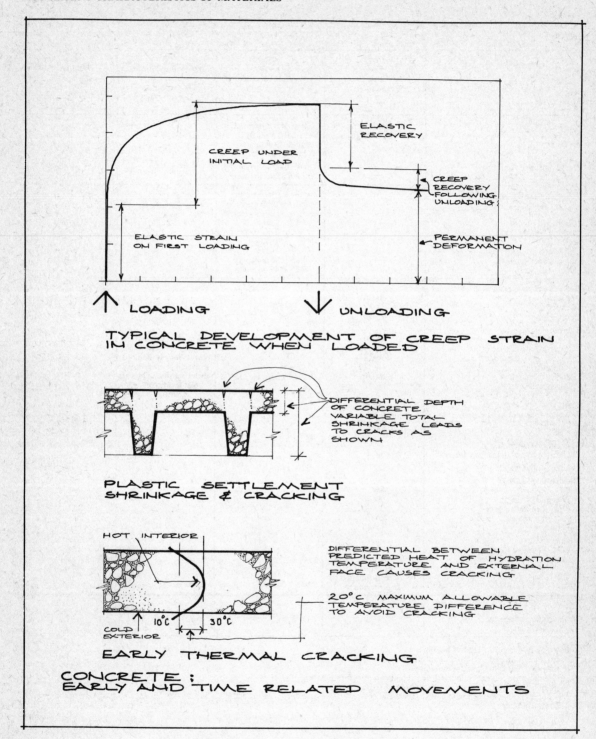

Diagram 3.1

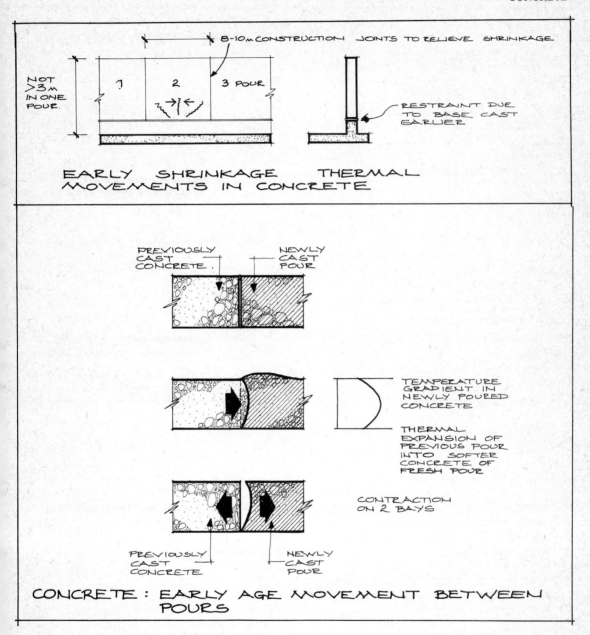

Diagram 3.2

MOVEMENT CHARACTERISTICS OF MATERIALS

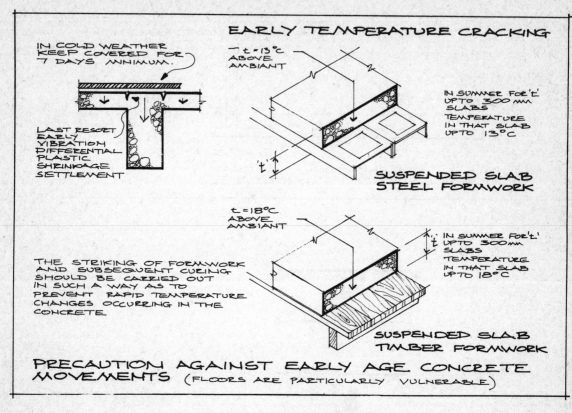

Diagram 3.3

PRECAUTIONS
1. Use of low heat of hydration cement.
2. Pour bays of 8–10 m with crack inducers.
3. Use smaller reinforcement at closer centres (eg 150 c to c)
4. Protect concrete from excessive differential eg in cold weather wood formwork to keep differential to interior to minimum.

Plastic settlement: movement

Due to settlement of solids in the plastic/liquid mix, leaving water termed bleeding water on top surface where it evaporates.

PRECAUTIONS
1. Low water/cement ratio
2. Prevention of premature drying out.

DEFORMATION MOVEMENTS IN THE HARDENED CONCRETE

See diagrams 3.4 and 3.5.

Causes

1. Elastic: due to loading (immediate on application of loads).
2. Thermal: due to temperature changes.
3. Alternate wetting/drying. It should be noted that concrete tends to loose water more slowly than the rate of absorption.
4. Long term initial drying shrinkage due to slow migration of initial mix water. This can continue for months.
5. Long term plastic deformation under load. This is called *creep*.

CONCRETE

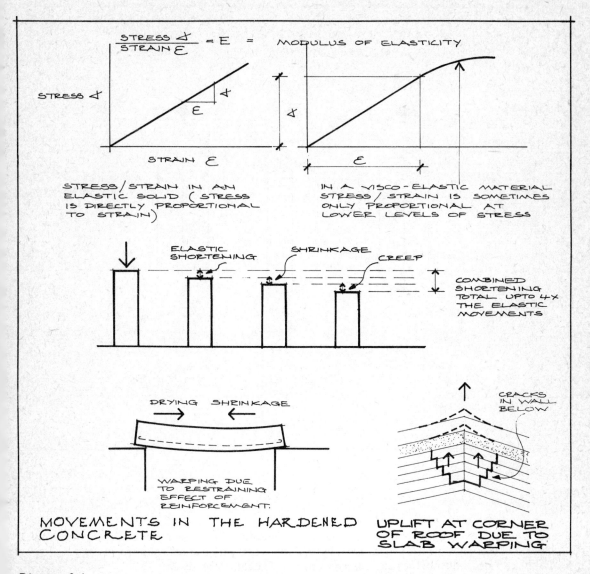

Diagram 3.4

1 Elastic deformation

The elastic modulus of concrete varies quite considerably with mix design and also type of aggregate. Values are given in CP. 110 for use in design. They vary from 25 KN/mm² for 20 N/mm² concrete to 36 KN/mm² for 60 N/mm² concrete.

2 Thermal movement

Again this varies with the volume and type of aggregate, water/cement ratio mix proportions and moisture content. The commonly accepted figures are:

Aggregate type	Co-efficient of linear thermal expansion $(\times 10-6)\,K-1$
(Quartzite) dense river gravel	10 – 14
Limestone	7 – 8
Lightweight	8 – 12

MOVEMENT CHARACTERISTICS OF MATERIALS

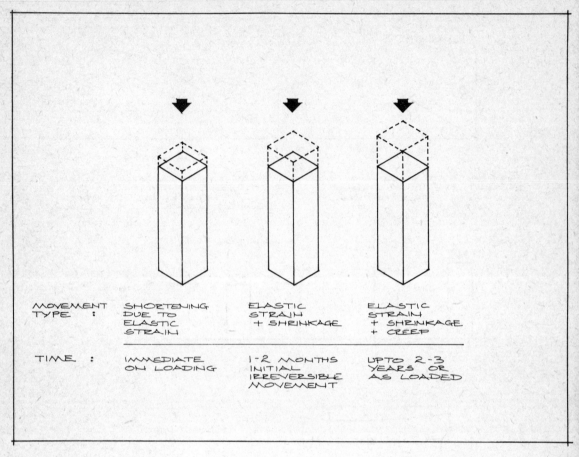

Diagram 3.5

The silica content of the aggregate has an influence on the rate of thermal expansion. Various rates of saturation can also influence the rate of expansion. For a full treatment of the thermal movements of concrete see the Current Practice sheet by the C and CA *Thermal movement of Concrete* by R D Browne where it is stated that the commonly accepted value of 12 micro strain/°C for partially saturated concrete with gravel aggregate is valid. Low saturation will produce higher values, eg in tension.

3 Drying shrinkage

Although this begins in the early age, ie it continues through setting and for a long period afterwards, its most serious effect is warping of slabs and curvature of columns.

The rate depends on:

1 Humidity of surrounding air
2 Rate of air flow
3 Ambient
4 Amount of water present.

In normal exterior conditions in Great Britain, half total shrinkage takes place in one month after casting, three quarters of total after six months and the rest over a longer period.

Variations of the amount of drying shrinkage in concrete.

Average (unreinforced) total 500×10^{-6}
(or .05%)
Low (granite or limestone
   aggregate)   0.045%
High (some shrinkable
   aggregates)   0.66 – 0.85%
(note values higher than .85% have been known)

Concrete with high shrinkage should not be used for this exposed reinforced section without air entrainment.

Drying shrinkage varies considerably in lightweight or aerated concretes. The following values of drying shrinkage are taken from BRE Digests 123 and 178.

Concrete type	Drying shrinkage %
Lightweight aggregates	
Pumice	0.04 – 0.08
Expanded perlite	0.2 – 0.3
Foamed slag	0.03 – 0.07
Expanded vermiculite	0.25 – 0.35
Aerated concrete varies with densities of 500 Kg/M³ – 880 Kg/M² (to B52028:1364:1963)	0.07 to 0.09

4 Creep

The amount of the creep depends on the following factors.

1 Age at which load is applied
2 Duration of loading
3 Relative humidity
4 Mix proportions and type of aggregate
5 Section geometry, ie surface area/volume

These factors are significant because creep is due to physical changes in the cement paste and to a far lesser extent on the type of aggregate.

5 Additive effect of creep and shrinkage

The exact computation of creep in concrete is beyond the scope of this volume, (see Concrete Society Technical Paper 16.101 *The Creep of Structural Concrete*) but the following example should show the serious nature of the final total movement.

Example 1 Vertical shortening of RC member 3M high (no allowance for reinforcement) 25 N/mm² 28 day strength, relative humidity of 65% compression 5N/mm² effective thickness 100×100 mm

		μ*	OU 3M
Shortening:	Drying shrinkage after 1 year	165	0.49
	Immediate elastic shortening	190	0.57
	Creep over 12 months	525	1.58
	Total shortening:	880	2.64

* ($\mu \triangleq$ micro strain = the strain $\times 10^{-6}$)

Prediction of deformation in the hardened concrete

(After L J Parrott: Simplified methods of predicting the deformation of structural concrete)

Precise methods now exist for the prediction of the deformations (inherent deviations) of structural concrete.

These have to take into account all factors which, as described earlier, have an influence on the extent, direction and type of deformation, eg the position of reinforcement, humidity, ambient temperature, loading, aggregate type etc. As curvature of members can result, this is obviously of concern to the Building Designer since the serviceability limits need to be defined, eg due to the proximity of glazing.

Prediction of this movement requires assessment of each of the following individual strains and their *combined effect*:

   Elastic movement   (E_{el})
   Creep   (E_{cr})
   Shrinkage (drying)   (E_{sh})
   Thermal   (E_{th})

Elastic strain (E_{el})

As described earlier, the modulus of elasticity of concrete varies with the concrete strength. This in turn increases over a period of time. Also to be considered are the elasticity of the aggregate, volume concentration of aggregate (ie mix) and cement paste.

MOVEMENT CHARACTERISTICS OF MATERIALS

The modulus of elasticity of concrete therefore increases from an initial value (E_o) as the concrete hardens and the designer has to calculate the Elastic Modulus of the concrete at the age of loading (E_t). This increase is proportional to the ratio of the strength at 28 days (f_{28}) and at the time of loading (f_t). (For the full equations see L J Parrot.) The final Elastic strain at the time of loading E_t can then be predicted.

Creep strain (E_{cr})

A simplified method in use today is the *creep factor* method. This enables the various factors (mentioned earlier) which influence creep behaviour to be combined in a simple equation:

Final creep strain $E_{cr} = E_{el} \times$ creep factor = $\dfrac{stress}{E_t} \times$ creep factor.

A typical chart for finding the creep factor is shown in the illustrations at the end of this section.

Shrinkage strain (E_{sh})

As described earlier, this movement depends on the relative humidity of the surrounding air and the proportion of surface exposed to it, as well as mix proportions, including water content. A simple graph for predicting shrinkage of concrete with 8% water content is shown in the illustrations at the end of this section.

The values of average shrinkage strain must be corrected if the reinforcement is asymmetrical and are also subject to seasonal variations of ± 0.4 times the 30 years shrinkage.

Thermal strain (E_{th})

As described earlier, the coefficient of linear thermal expansion of concrete varies with the aggregate type and the degree of saturation of the cement paste, partially dry concrete has a coefficient approximately $2 \times 10^{-6}/°C$ higher then saturated concrete. After finding this coefficient, the actual amount of movement is then calculated as described earlier, by multiplying the coefficient by the expected variation in service temperature and size of member.

A chart showing the variation of coefficients of thermal expansion for concrete of different aggregate types is shown on the facing page.

CONCRETE BLOCKWORK

The properties of masonry blockwork are similar to those of the concrete type used in the blocks. These properties are as given earlier for concrete.

The properties are modified by the mortar constituent. Creep is not so significant as in *in situ* concrete because the fully matured blocks only are loaded.

Initial drying shrinkage is, however, still proceeding after installation, particularly of 'green' blocks, ie recently cast.

Elastic deformations

Creep modified E values are 1–3 Kn/mm^2 for lightweight blocks and 2–6 Kn/mm^2 (dense blocks).

Thermal movement

5.6–$9.4 \times 10^{-6}/$deg C^{-1} depending on block type.

Drying shrinkage: see properties of constituent concrete type.

3.4 BRICKS AND BRICKWORK

Brickwork, being a composite of bricks and mortar naturally – derives its properties from these two constituents.

Early age movements: irreversible

This is mainly expansion due to the take up of moisture after their kiln dry state in clay bricks and initial shrinkage of concrete or calcium silicate brickwork in curing. These movements take place mainly in the first few weeks after manufacture. Diagram 3.6 shows the differential movements which may occur in various types of brickwork.

Loading: elastic deflection and creep

The elastic deflection varies considerably with the strength of brickwork and the mortar mix. Creep

BRICKS AND BRICKWORK

Coefficient of Thermal Expansion

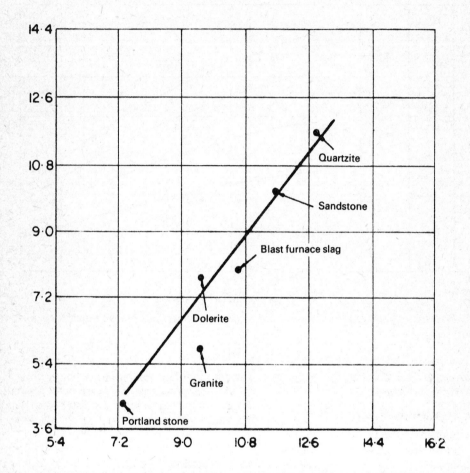

Coefficient of thermal expansion of concrete × 10^6 per degree centigrade

Copyright from Bonnell & Harper
National Building Study: Technical Paper No. 7

is not as significant as in concrete. The elastic modulus can vary from 5 Kn/mm² for flettons in 1:¼:3 mortar to ₂0.0 Kn/mm² to 35 Kn/mm² for high strength bricks.*

Nevertheless, the 'creep modified' (ie final movement) modulus is 25–50% higher than the initial elastic modulus.

* LENTZNER, D, *Movement in Buildings*.

MOVEMENT CHARACTERISTICS OF MATERIALS

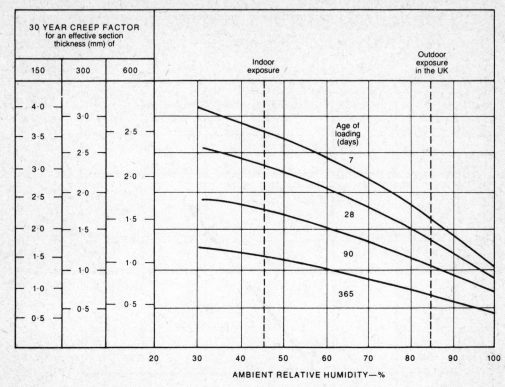

Effects of relative humidity, age of loading and section thickness upon creep factor[6] for concrete

1 Thermal movement (see Co-efficients of linear expansion, pages 26 and 27)

Note this varies as follows +

Clay bricks +	highest (vertical)	–	8–12
	width (less joints!)	–	4–8
	length	–	4–8
Calcium Silicate	height		14–22
	width		14–22
	length		11–15

+(Fion BS 6093:1981)

The average figure given for horizontal thermal movement for most brickwork is 5.6×10^{-6} and for vertical dimensions $(5.6 \times 1.5) \times 10^{-6}$

2 Sulphate action

This is due to the expansion of jointing mortar due to the chemical action of sulphates on the cement. The amount of movement cannot be predicted and should be prevented by correct specification/design, mainly to prevent excessive saturation of the brickwork.

3.5 TIMBER

Wood is a highly unisotropic material and most of its movement characteristics vary with

1 Species
2 Direction of grain (tangential or radial)
3 The moisture content.

Variations in ambient relative humidity cause moisture content changes and movement in timber are mainly due to this.

Moisture movement in timber

Table 3.5 gives the movement of timber in the normal range of moisture movement, of 10% to

BRICKS AND BRICKWORK

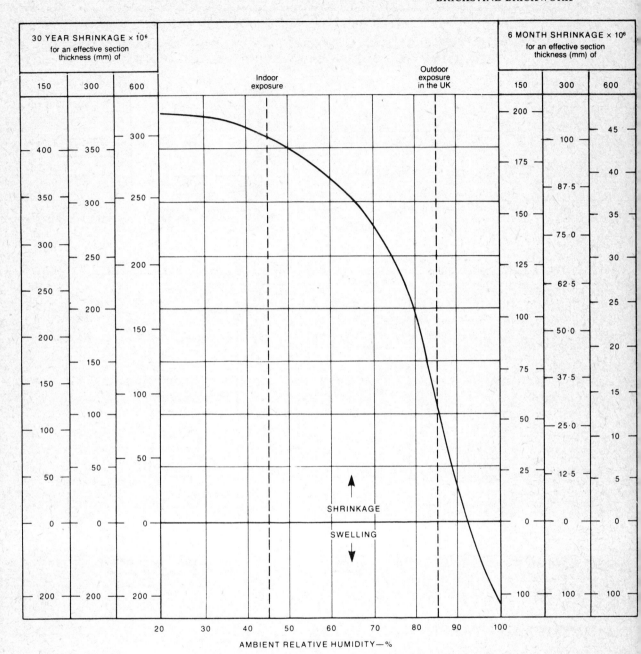

Drying shrinkage of structural concrete

MOVEMENT CHARACTERISTICS OF MATERIALS

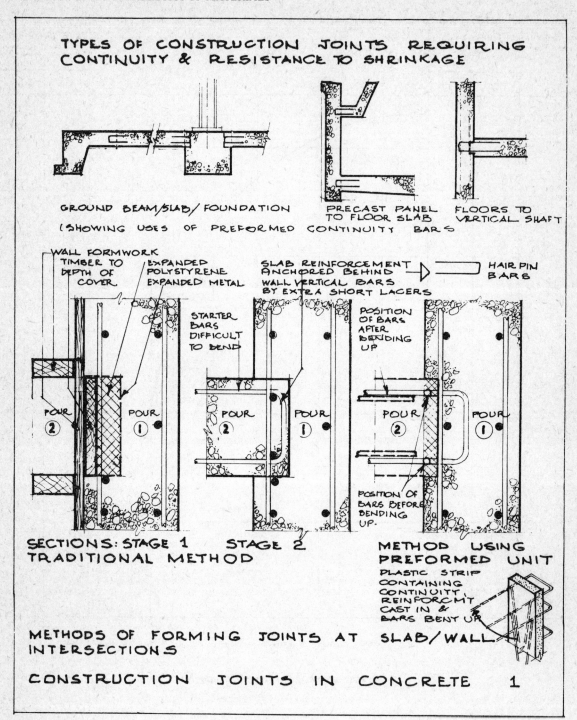

For fuller treatment of joints in concrete structures see pages 107 to 110.

LOCATION OF JOINTS

VERTICAL EDGE TO COLUMN CAP KEPT TO 75 MM MINIMUM TO ALLOW LEVELLING OF SLAB

SLAB — JOINT 12mm INTO SLAB — COLUMN

SECTION OF COLUMN/SLAB

SLAB — JOINT 12mm INTO SLAB — BEAM

SECTION OF DEEP BEAM/SLAB

STRIATED SMOOTH STRIATED FINISH. BAND FINISH — ③ KICKER ② ①

SMOOTH BAND — ③ KICKER TO LOCATE WALL SHUTTERS ② SMOOTH BAND ① **TROUGH SLAB/WALL**

SECTIONS: SLAB/WALL JUNCTIONS

LOCATION OF FEATURE BANDS ON FACE HIDE CONSTRUCTION JOINT LINES AND GIVE TOLERANCE IN CASE OF DEVIATION IN FACES

CONSTRUCTION JOINTS IN CONCRETE 2

MOVEMENT CHARACTERISTICS OF MATERIALS

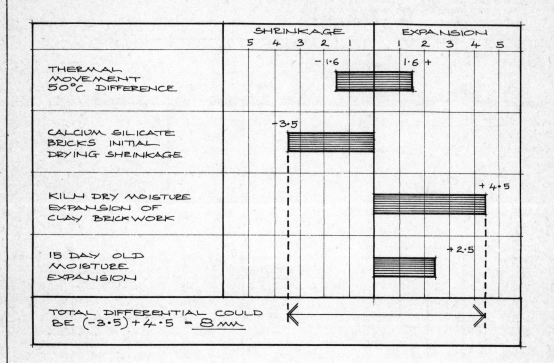

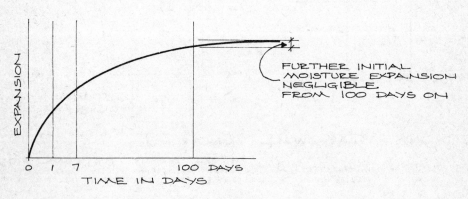

Diagram 3.6

18% in buildings. These variations occur when the ambient pH at 25°C is 60–90%

Table 3.5

Species	Moisture content variation	Corresponding movements %	
		Tangential %	Radial %
Western Hemlock	21–13	0.9	1.9
Scots Pine (Redwood)	20–12	2.1	0.9
European Spruce	20–12	2.1	1.0
Teak	15–10	1.2	0.7
Oak (English)	20–12	2.5	1.5

Thermal movement of Timber

Again this varies in direction:

across grain: 50 to $60 \times 10^{-6}/C^{\circ -1}$
with grain: 3.8 to $6.5 \times 10^{-6}/C^{\circ -1}$

(along the length therefore this is similar to the movement of brickwork!)

Elastic movement of timber

This also varies in direction and by species as follows:

Softwood

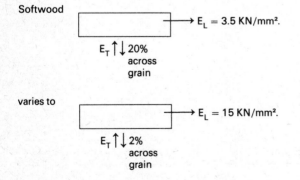

varies to

and Hardwood

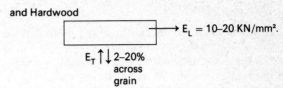

E_T = Elastic modules: transverse grain.
E_L = Elastic modules: along grain
ie E_T can be $\frac{1}{20}$ of E_L.

3.6 CERAMICS

Ceramic wall/floor tiling

Although fired clay products in general can be classified under this heading, this section summarises the movement characteristics of ceramic wall tiling.

Glazed and ceramic tiles

Seemingly inert and immobile, yet bulging and cracking frequently occur. Glazed tiles have a porous/biscuit body which renders them more prone to moisture movement than vitrified tiles. Clay tiling is less elastic than concrete or brickwork so for the *same* movement more stress is induced leading to cracks. Similarly the (thermal) co-efficient of linear expansion, α is $4-6 \times 10^{-6}/C^{\circ}$ as against 5–8 of normal brickwork.

Unfortunately tiling is subjected to shrinkage strains of the backings: rendering or floor screeds and these are higher than tiles and joints can withstand, *unless* the strain is limited over small areas, hence the need for movement joints in tiling.

Movement joints should be 6 mm wide, spaced 3 m horizontally and 4.5 m vertically and over *all* movement joints in the substrate.

3.7 METALS

Movement characteristics

Metals vary greatly in their movement characteristics.

Reference should be made to the table at the beginning of this section.

Particular note should be taken of the following:

MOVEMENT CHARACTERISTICS OF MATERIALS

Elasticity under load

Most metals are truly elastic within usual service loads. Note that the difference of 70 to 210 Kn/mm² for the modules of Elasticity of Aluminium to steel ie for same section and/or same load aluminium deflects three times as much as steel.

Thermal expansion

Mild steel and stainless steel vary 21.1 to $17.3 \times 10^{-6}/°C$. The non ferrous metals, traditionally moving considerably, especially when on roof coverings in conditions of maximum exposure have a co-efficient of linear expansion of 28.6 (lead) and 17 ($\times 10^{-6}/°C$ for copper.

3.8 PLASTICS

Properties

These vary considerably by composition. However as a general rule, Thermal movement is far greater than that for metals and ceramics (clay brickwork) so that plastics used for windows, panel cladding, etc, will be subject to large movement differentials which must be allowed for. The elasticity of plastics is well known but also varies greatly according to composition. See table 3.3.

Thermal movements

In addition to the BRE guide given in table 3.3 quoted earlier from BRE Digest 228, the following summary of co-efficients of linear thermal expansions is given by BS 6093: 1981 may be useful.

Acrylic – cast sheet	$50–90 \times 10^{-6}/°C$	
Polycarbonate	60–70	
Polyester 30% glass (BRP) fibre:	18–25	(varies if fibres directional)
Rigid PVC	42–72	
Phenolic	30–45	
* Expanded polystyrene	15–45	
* Foamed polyurethane	20–70	
* Foamed phenolic	20–40	
* Expanded PVC	35–50	

* *Note:* When used as roof insulation, maximum service temperature fluctuations can be expected.

4 Design for movement in buildings

4.1 RELATIVE MOVEMENT PROBLEMS

Movement in buildings can originate in the structural elements and be transmitted to the non-structural elements or visa versa. Either can lead to failures in the other. One element may be restrained from moving by another and their relative stiffness provides the restraining force. This interaction between all elements needs to be taken into account by all concerned at the design stage and a 'strategy' for the building agreed.

ALTERNATIVE STRATEGIES

In the control of movements each building will present a unique problem, but several broad strategies are possible. These have been used in detail in the guides for element design in section 5, as follows:

(a) *Suppression of movements* (translations into stress)
By reinforcement, friction, barriers, controlled cracking. Where movement is restrained, the stresses resulting must be accommodated and not exceed the acceptable limits for the material.
This alternative is used in reinforced concrete only to a limited extent and in finishes over small areas.

(b) *Prevention/reduction of movements*
by design features articulation and hinging, insulation.
Protection: reflective colours. By construction methods/choice of materials:

Mix design in concrete
Choice of construction time/protection/curing.

This alternative is covered in detail in the appropriate sections.

(c) *Accommodation of movements* (allowing free or partially free movement to take place and to design modifications to allow elements to accept movements.
By provision of separation and/or provision of movement joints/slip planes/allow limited dislocation (controlled cracking). Each of these alternatives may be employed in a single project.

(d) *Movement accommodation methods*
This alternative can allow for the following variations:

1 Free movement (of whole components or elements)
2 Restraint (at local level)
3 Partial restraint (in one location or direction), ie leaving component free to move in a limited direction or plane only.

Fundamental principles in design for movement accommodation

In order to decide upon the correct strategy or combination, it is essential to consider both the *form* of the building, the *materials* of which it is composed, the *environment* to which it will be subjected *during erection* and *in use*.

The following factors need to be considered:

1 Physical properties of materials of construction
2 All potential movements, ie changes of size.
3 All restraints: to assess the stresses resulting from degree of restraint provided.
4 Assessment of variability of conditions or materials.
5 Fixings, bearings, loadbearing mechanisms must continue to function
6 and movement must continue to be accommodated in the life of the building.
7 Any internal restraints in components/materials otherwise free to deform.

RELATED DESIGN PROBLEMS

At movement joints

These are required to perform in other aspects than movement, particularly *weather resistance* (see BS 6093: 1981 for a complete list of joint functions).

The joint must accommodate movement but also inaccuracy in components and erection.

Fatigue due to frequent movement is also possible and noise problems.

At fixings

Adequate adjustments to allow for movement in the appropriate direction and inaccuracies needs to be provided.

ASSESSMENT OF MOVEMENTS AND INACCURACIES

In joint design a realistic assessment of inaccuracies as well as movements is required.

Trends in movement distribution

It is also necessary to recognise the following effects in the distribution of movements in a building:

1 Stationary/fixed planes: At rigid parts of the building, movement will be *zero*, ie directed away from these planes.
2 'Fan' effect: The factors causing movement may vary at a joint plane so this may move unequally along its extent.

Mode of movement

This can be slow or sudden, repetitive, frequent/infrequent. Small displacements of some elements *may not be reversible* leading to permanent displacement with serious consequences. This has been termed 'shuttle' effect.

4.2 DESIGN FOR MOVEMENT: SUMMARY OF PROCEDURE

The following is a summary of the stages necessary in the design for movement control:

1 Recognise and assess environmental factors (see section 2)
2 Assess extent of problem, including movement 'trends', building size
    shape/storey heights/solid-voids/spans, materials (see diagram 4.1)
    foundation type
    subsoil (see section 5.1)
3 Identify problem areas:
    points of restraint.
    change of form/stress, differentials (see section 5.1 for elements).
4 Assess or estimate movements due to various causes (see section 2).
5 Decide on limiting factors and failure limits (see section 5.1 and tables).
6 Decide on strategy to be adopted.
7 Incorporate design features for chosen strategy, ie the following alternatives:
   (a) reduction/prevention
   (b) suppression: partial or complete restraint
   (c) movement accommodation either limited dislocation or *place and* design. Movement joints/slip planes (see 4.3) *or complete separation/articulation* (see section 5.1).

4.3 MOVEMENT ACCOMMODATION

Movements can be accommodated in various ways, but a discontinuity in the fabric is usually required to allow free movement without restraint.

General need for joints: joint types

The provision and location of joints in buildings arises from the assembly requirements of the materials and components and their junctions, ie practical constraints.

Only some of these joints will have to be designed to accommodate movements.

Joint types: functional requirements

It is essential to identify the differences between various types of joint so that the special requirements of movement joints can be considered. The definitions are taken from BS 6093: 1981.

(a) *Non-movement joints*
    These are either *type 1:* No allowance for deviations.

MOVEMENT ACCOMMODATION

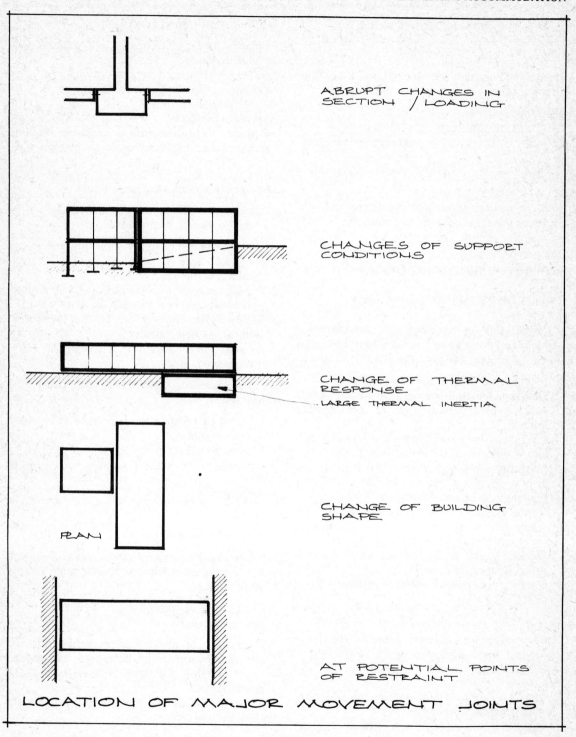

Diagram 4.1

DESIGN FOR MOVEMENT IN BUILDINGS

　　　　　or　　type 4: Allowance for *induced* deviation only

(*induced* deviations are due to work, eg site/manufacturing jointing tolerances)

(b) *Movement joints*
　　These are either *type 2*: allowance for induced and *inherent* deviations
　　　　　or　　*type 3*: inherent deviations only.

(*Inherent* deviations are due to inherent changes, eg stress, temperature, moisture changes)

4.4　MOVEMENT JOINT DESIGN

PROCEDURE IN JOINT DESIGN

After deciding on movement accommodation as strategy and deciding to provide movement joints. The following outline procedure can be adopted:

1　Evaluate movements due to all causes.
2　Decide on location/spacing of movement joints.
3　Adapt general fabric design to accommodate joints.
4　Check that fixings can accommodate movement and leave movement joints free to move.
5　Establish joint performance criteria, eg exposure, durability, environmental factors.
6　Establish dimensional factors:
　　induced deviations – tolerances and fits calculate maximum inherent deviations, ie calculate maximum movements at joints.
7　Decide on joint mechanism and geometry profile/shape/width.
8　Choose appropriate method of sealing.

Note　The dimensional/sealing requirements may lead to reconsideration of spacing and joint width.

This procedure follows in principle the design procedure for joint design in BS 6093: 1981 to which reference should be made for a fuller treatment.

The following guidance notes are intended to be used in this design procedure.

Detailed guidance is given for the appropriate elements/materials in section 5. See also diagram 4.1. However the following locations are likely:

Location of movement joints

At planes of change in construction material
At planes of shape
At changes of thermal response/capacity
At changes of loading or loading capacity
At changes of direction
At points of restraint
At planes coinciding with a hiatus in building operation or natural junctions between elements.

Movement joint types and performances

In the design of joints it is essential to identify the accommodation to be provided.

1　*Contraction* (Shrinkage) joint for partial or complete separation. Accommodates movement in *one* direction only.
2　*Expansion joint*: movement in two/three direction or rotation construction discontinuous.
3　*Sliding joint* movement in one direction support/friction provided.
4　*Hinged joint*: rotation, construction continuous only.

Other performance requirements to be considered where appropriate.

1　Weather proof/air/moisture vapour resistance.
2　Fire resistance/.
3　Durability adhesion and integrity of sealant.
4　Feasibility of formation/sealing.
5　Freedom of movement.
6　Appearance.

Movement jointing methods

In order to achieve a satisfactory joint, which will give the required performances, the main factors to be decided are.

1　Joint profile/shape and width (see diagram 4.2)
2　Sealing method.

The method of sealing can influence the profile and certainly the width and both aspects will depend on the direction and type of movement to be accommodated.

Diagrams 4.3 and 4.4 show various profiles and sealing methods for

(a) single skin one stage joints.
(b) single skin two stage joints.

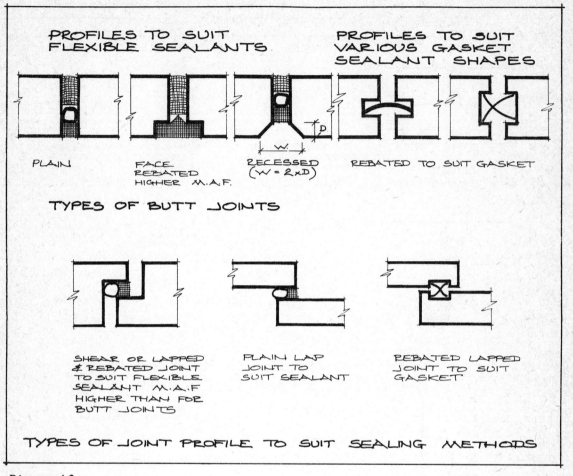

Diagram 4.2

Selection of joint profiles

In selecting a jointing method the need for ease of application, durability should be considered as of paramount importance.

The face of the component to be joined must be taken into consideration in deciding on joint profile and depth (see diagram 4.5).

The selection of joint profile in sealed joints should take account of the advantage of protection afforded by shear joints as against butt joints (see diagram 4.6) due to the protection of the sealant.

CALCULATION OF MOVEMENT JOINT WIDTHS

In butt type joints and sealed joints generally, the width of the joint is vital to the integrity of the sealant, as well as the general fit of the components being joined.

The joint widths are calculated in *two stages*

1 The minimum joint width to accommodate the movement

This depends on:

(a) The maximum anticipated or total movement at the joint: TM
(b) the movement capability of the sealant, expressed as a percentage of   MAF
movement over original size and gives the minimum joint width   Cd

The following relationship gives the minimum

DESIGN FOR MOVEMENT IN BUILDINGS

SINGLE SKIN ONE STAGE JOINTS

	SHAPE	TYPE	SEALING METHOD
VERTICAL JOINTS	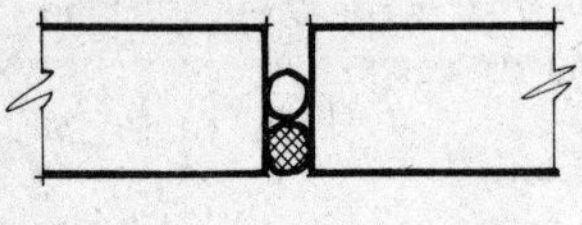	BUTT SEALANT FACED JOINT	FLEXIBLE SEALANT WITH BACK UP STRIP
	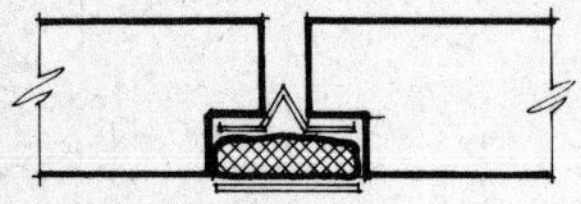	RECESSED OR REBATED WITH BONDED MEMBRANE + COVER STRIP	BONDED MEMBRANE SEALANT & COVER STRIP
	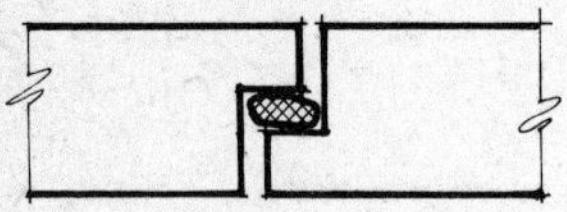	LAPPED JOINT	SEAL OR GASKET INSERTED DURING ASSEMBLY
		PROFILED SHEETS	LAP ONLY - LABYRINTH ACTION : SEALANT OPTIONAL
HORIZONTAL JOINTS	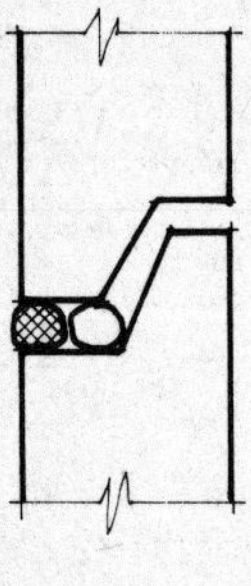	BUTT SEALANT FACED	FLEXIBLE GUN APPLIED SEALANT
	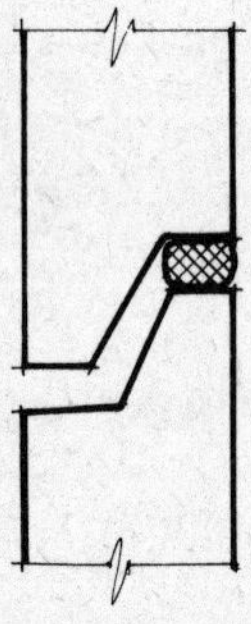	RECESSED OPEN DRAINED JOINT	PREFORMED ADHESIVE STRIP, FLEXIBLE SEALANT OR GASKET

Diagram 4.3

MOVEMENT JOINT DESIGN

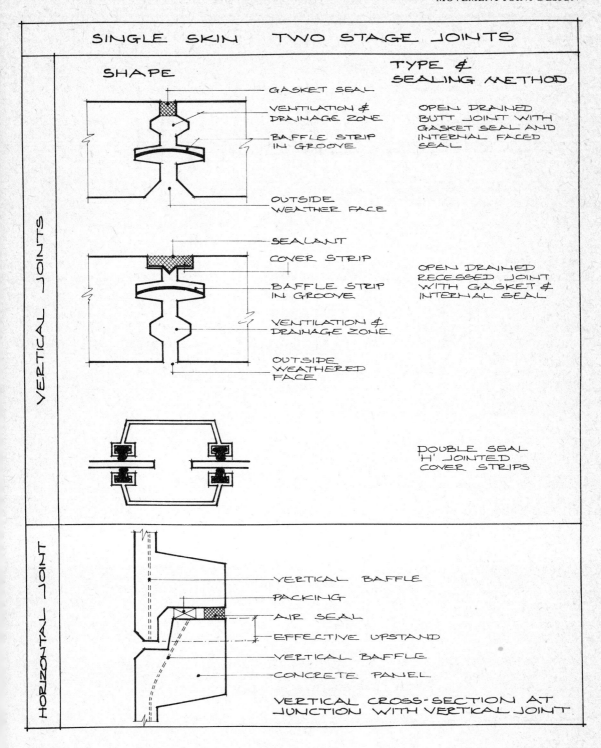

Diagram 4.4

DESIGN FOR MOVEMENT IN BUILDINGS

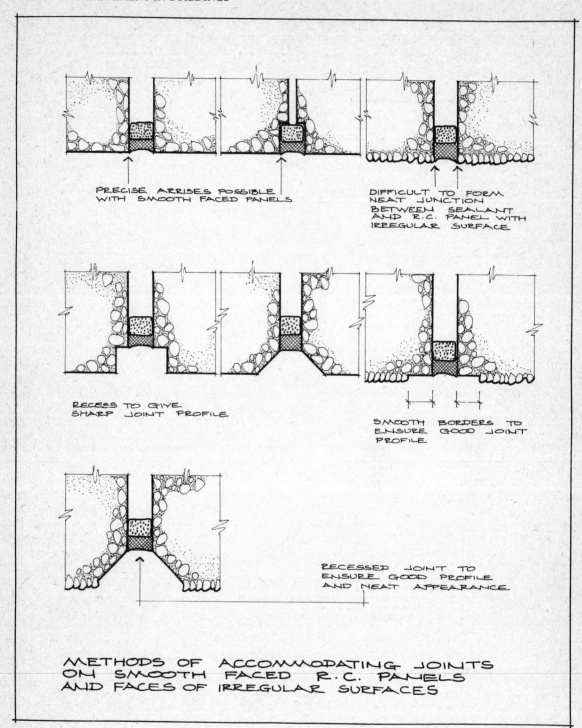

Diagram 4.5

MOVEMENT JOINT DESIGN

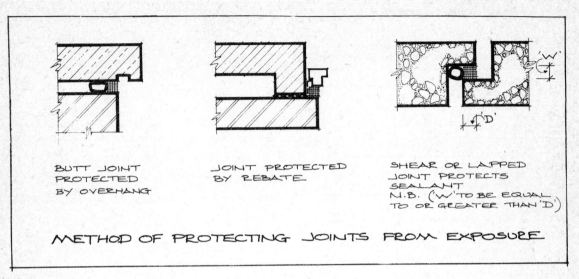

Diagram 4.6

joint width to ensure the sealant remains within its movement capability even if sealed in extreme conditions (cold or heat).

$$Cd = \frac{TM \times 100}{MAF} + TM \text{ mm}$$

for example, for a total movement TM of 4 mm and a MAF of 25%,

$$CD = \frac{4 \times 100}{25} + 4 = 20 \text{ mm}.$$

2 Calculation of design joint width

This must take account of all (induced) deviations due to inaccuracies in manufacture and erection and within the maximum and minimum joint widths. (It must also take into account the practical maximum and minimum of 6 mm minimum for filling, 50 mm maximum for slump. If Cd exceeds this, use more joints.)

Allowance for deviation and fit

The allowance for deviations and fit of components can be derived by use of graphical aids for tolerances and fits HMSO London. See also BRE Digest 199 March 1977 – getting a good fit. See diagrams 4.7, 4.8 and 4.9 for a graphic illustration of the principle of tolerances imposed on normal distributions of inaccuracies in size/position.

Since allowance must be made for expected deviations within the calculated maximum and minimum joint width in addition to the movement to be accommodated it is essential to decide on a realistic level of tolerances for the building at an early stage.

BS 5606: 1978 *Code of Practice for accuracy in Building*, gives three tables showing deviations from the specified dimensions for various types of construction.

Table 1 of the Code gives the average deviations of the measured sizes from those specified, which were found by the BRE survey, in two aspects:

1 displacement of mean, ie by how far the average sizes differed from those specified.
2 the spread of sizes found (standard deviation).

Note Each kind of construction was found to have its own 'characteristic accuracy' or spread.

The designer is asked to assess how significant the effect of a 'displacement of mean' (sizes) is when calculating for fit.

Note Table 1 also draws a distinction between the data which will control 'fit' ie the ability of components to fit into *spaces* between elements, and data primarily included for the control of accuracy in the *sizes* of components or elements.

Section 3 of the Code gives guidance to the

DESIGN FOR MOVEMENT IN BUILDINGS

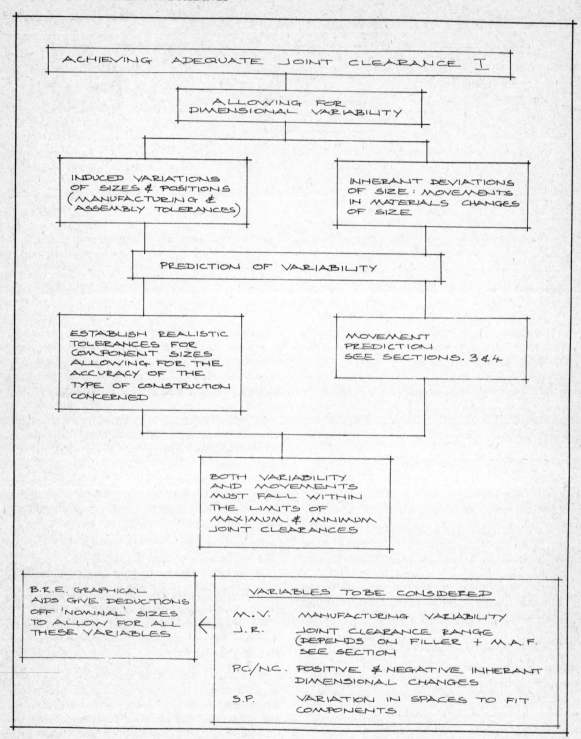

Diagram 4.7

MOVEMENT JOINT DESIGN

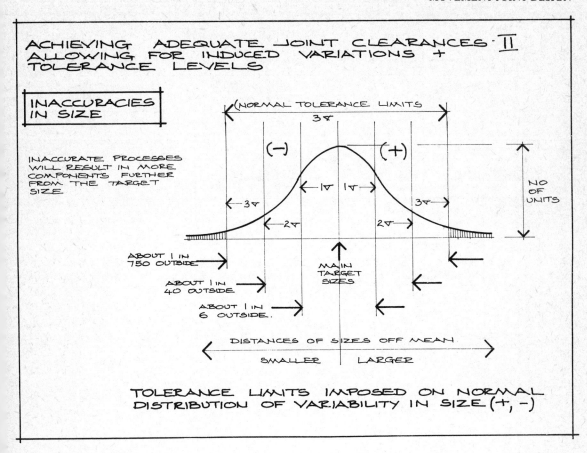

Diagram 4.8

designer in specifying permissible deviations. The procedure allows for 'inherent deviations' ie movements and 'induced deviations' ie departures from the intended *size* and *position* of components or spaces due to setting out, manufacturing, assembly.

Tables 2 and 3 of the Code give estimated values of deviations for various types of construction to guide the designer in specifying realistic permissible deviations for any particular construction system

The following extract, table 4.1 gives some of the values given in table 2 of BS 5606: 1978. For full guidance reference should be made to the tables in Section 3 of the Code.

Joint depth

In butt joints with elastic sealants this should be half the width.

Joint sealing mechanisms have been illustrated in diagram 4.2.

The following components in joint formation are usually required (see diagram 4.9).

1 Fillers: used to close the lower depth of joint leaving space for sealant.
2 Bond breakers: Separating strips to allow sealant freedom to move without adhesion to the joint filler.
3 Back up material: to control depth of sealant.

The performance requirements of a sealant are:

1 to accommodate movement without deterioration or loss of adhesion
2 provide a seal against rain, snow, air, dust
3 exclude chemical/biological agents
4 provide sound/fire resistance.

Table 4.1 Permissible deviations in building based on measured characteristic accuracies (based on Table 2 of BS5606:1978)
All values are expressed in millimetres, na means not applicable

Item of construction		Type of dimension measured	Brickwork	Blockwork	In situ concrete	Precast concrete	Steel	Timber
Space between elements	Between walls	at floor[1]	±20	±20	±30	±20	na	±40
		at soffit[1]	±25	±30	±30	±25	na	±45
	Between Columns	at floor[1]	na	na	±20	±15	±15	—
		at soffit[1]	na	na	±25	±15	±15	—
		Cased steel at floor[1]	na	na	na	na	±20	na
		at soffit[1]	na	na	na	na	±20	na
	Beams Floor slabs	Floor to soffit height	na	na	±30	±25	—	—
Openings	Window or door	Width up to 3 m (not jigged)	±25	—	±20	±15	na	—
		Height up to 3 m (not jigged)	±25	—	±25	±15	na	—
Size and shape of elements and components	Walls	Height up to 3 m	±35	±40	—	—	na	±20
		Thickness	±25	—	—	—	na	na
		Straightness in 5 m	±7	±8	±10	±9	na	—
		Verticality up to 2 m	±10	±10	±15	±10	na	—
		up to 3 m	±10	—	±20	±15	na	—
		Level of bed joints 3 m	±15	±15	na	na	na	—
	Columns	Size up to 1 m	na	na	±10	±8	—	—
		Verticality up to 3 m	na	na	±15	±10	±8	—
		up to 7 m	na	na	—	—	±10	—
		Cased steel up to 3 m	na	na	—	—	±10	—
		Squareness	na	na	±10	±6	na	—
	Beams	Depth (perimeter beams)						
		up to 600 mm	na	na	±15	—	na	—
		over 600 mm	na	na	±25	—	na	—
		(internal beams)						
		up to 600 mm	na	na	±15	—	na	—
		over 600 mm	na	na	±20	—	na	—
		Level[2] [3] flatness	na	na	±25	±20	±10	—
		variation in datum	na	na	±15	±25	±20	—

MOVEMENT JOINT DESIGN

Item of construction		Type of dimension measured	Brickwork	Blockwork	In situ concrete	Precast concrete	Steel	Timber
Structural floor		Level[3] (based on a 2.5 m grid) flatness	na	na	±20	±20	na	—
		variation in datum	na	na	±25	±30	na	—
		Precast with in situ topping Level[3] (based on a 2.5 m grid) flatness	na	na	±20	na	na	na
		variation in datum	na	na	±35	na	na	na
Screeded floor		Level[3] (based on a 1 m grid) flatness	na	na	±6	—	na	—
Structural soffit		Level[3] (based on a 2.5 m grid) flatness	na	na	±20	±20	na	—
		variation in datum	na	na	±15	±7	na	—
Panels		Length	na	na	—	—	na	±6
		Height	na	na	—	—	na	±6
Overall size (on plan)	Building	Length or width up to 40 m	±40	—	±35	±55	±20	—
	Ground floor slab	Length or width	na	na	±40	—	na	na

[1] These dimensions are horizontal measurements taken at the levels indicated (see Appendix A of the BS Code for further details)
[2] The flatness and variation in datum of beams is measured on the soffit of concrete beams but on the top of steel beams.
[3] The deviations from intended level of beams, floors and soffits have been separated into two parts:
    'flatness', a measure of the variability in level from the average plane;
    'variation in datum', a measure of the variability in level of the average plane from its intended level.

Note To obtain the total permissible deviation of combined 'flatness' and variation in datum, the recommended technique given by the Code is to find the total standard deviations being the sum of the separate deviations due to various causes which are *additive*, the total displacement of mean 'M' and combine these to find the total permissible deviation = ± (35+M)

55

DESIGN FOR MOVEMENT IN BUILDINGS

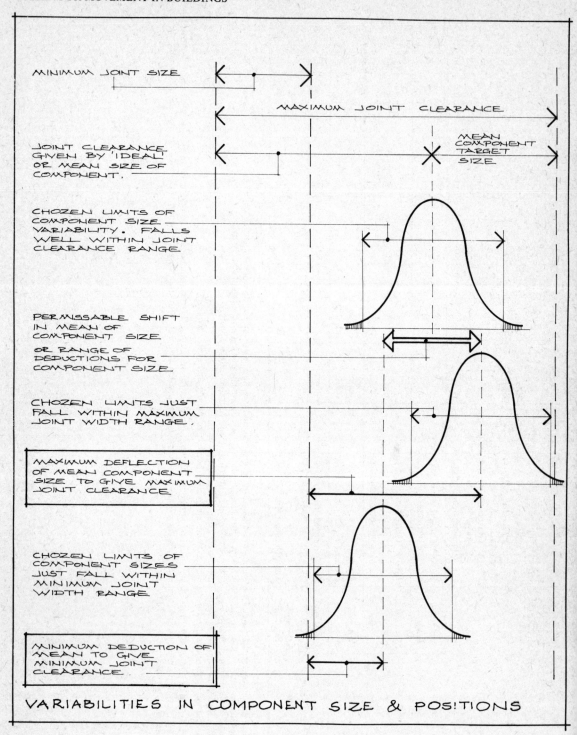

Diagram 4.9

MOVEMENT JOINT DESIGN

Selection of sealants and fillers

This often depends on the feasability of application, the materials at the joint and the durability required.

The following tables give basic data for the selection of suitable joint fillers and sealants.

Table 4.2 (taken from BS 6093/1981 table 4) gives the classification of joint seals and fillers.

Table 4.3 (taken from BS 6093/1981 table 5) gives a list of suitable joint fillers.

Tables 4.4 and 4.5 (taken from Manual of Good Practice in Sealant Application Table 1 (SMC/CITIA) and DOE Construction Guide 18) gives guidance in selection of a suitable sealant for a given joint situation with data on life expectancy and movement accommodation factor.

Table 4.6 gives a guide to the joint width required for various total movements, based on the (MAF) movement accommodation factor of the sealant used.

The joint profile will also influence the choice of joint sealing method (see diagram 4.2).

The main types are:

1 *Preformed gaskets* which are either sprung into position or compressed, they rely on the compression to be maintained. They cannot accommodate irregularities or damaged joint edges. Gaskets can be cellular or non cellular, ie profiled.
2 *Flexible sealants*
  The following chart classified the types available taken from the Manual of good practice in sealant application published by CIRIA and the sealant manufacturers conference January 1976.

For guidance in the selection of constructional sealants the following charts which are reproduced for reference here (diagram 4.2 and 4.6) may prove useful. They are reproduced from the PSA advisory leaflets on flexible sealing materials. A draft British Standard on sealant selection should also be available.

JOINT DESIGN AND DETAILS

Other functional requirements of the joint

This volume deals primarily with the design of joints in order to accommodate movements. In the design of joints for the external building envelope the problem of the preservation of the weatherproof skin is of primary importance.

Jointing between cladding panels as illustrated earlier (one/two stage sealed) has to a large extent been rendered less critical by the 'rain screen' principle, where there is a ventilated drained space behind the panels. However, major structural movement joints will always be potential sources of problems. The fire/acoustic resistance of dividing elements is also affected (see diagram 4.11) for the fire endurance test result on a one stage butt joint). BS 6093/1981 gives a checklist of joint functions which require to be taken into consideration, and is reproduced in table 4.7.

Joint formation

The theory and design of jointing is incomplete without taking into account the practical problems of formation and careful workmanship on site.

The more complex the joint profile or difficult the method of formation, the greater is the likelihood of failure.

Gasket joints are particularly vulnerable to failure if they rely on a sharp profile, particularly at intersections where corners of units are chipped in transit.

Concrete edges of joint profiles must be carefully inspected after striking the shuttering when damage is frequently caused to the joint profile.

Diagram 4.12 shows the difficulty in placing 'internal' water stops and the comparative ease of placing 'external' water stops in shrinkage joints in *in-situ* reinforced concrete work.

Finally, the instructions by sealant suppliers for priming the joint edges should be carefully observed where applicable, and, if possible, the work of sealing all joints, even simple/wall joints, should be carried out by a reputable specialist subcontractor.

Table 4.2 Classification of seals and fillers

Classification	Joint seals			Joint fillers
	Sealants (flexible)	Sealing strips	Gaskets	
Product properties	Cohesion, adhesion, elasticity, plasticity	Cohesion, adhesion, elasticity plasticity	Cohesion, compressibility	Surface keying
Material form	Unformed gun, pourable or knife grades	Cellular and non-cellular strips and coated gaskets	Cellular and non-cellular in solid, hollow or finned sections or combinations	Cellular, fibrous or granular in sheet, strip or unformed
Material types	Polysulphide: one-part Polysulphide: two-part Silicone: one-part Polyurethane: one-part Polyurethane: two-part Acrylic: one-part Butyl Oil based Bituminous Bitumen rubber Pitch polymer	Extrusions based on sealant materials ranging in consistency and in properties from plastic to elastic	Neoprene Natural rubber Ethylene propylene diene monomer Butyl Polyurethane Polyethylene Ethylene vinyl acetate Silicone Cork Mineral fibre	Wood fibre/bitumen Bitumen/cork Cork/resin Cellular plastics Cellular rubbers Mineral fibre Synthetic fibre Mortar

Table 4.3 Fillers for movement joints

Joint filler type	Property					
	Typical uses	Form	Density range	Pressure for 50% compression	Resilience (recovery after compression)	Tolerance to water immersion
			kg/m³	N/mm²	%	
Wood fibre/bitumen	General purpose expansion joints	Sheet, strip	200 to 400	0.7 to 5.2	70 to 85	Suitable if infrequent
Bitumen/cork	General purpose expansion joints	Sheet	500 to 600	0.7 to 5.2	70 to 80	Suitable
Cork/resin	Expansion joints in water-retaining structures where bitumen is not acceptable	Sheet, strip	200 to 300	0.5 to 3.4	85 to 95	Suitable
Cellular plastics and rubbers	Low load transfer joints	Sheet, strip	40 to 60	0.07 to 0.34	85 to 95	Suitable if infrequent
Mineral or synthetic fibres	Fire-resistant joints: low movement	Loose fibre or braided	Dependent upon degree of compaction	Dependent upon degree of compaction	Slight	Not suitable

Tables 4.2 and 4.3 based on Table 4 BS 6093:1981

Table 4.4 Properties of sealants

Sealant	Chemical type	*Nature	Life expectancy (years)	Movement accommodation %	Max. joint width (mm)	Comment
Hot poured	Bitumen	Plastic	3–10	5–10	50	Available to BS 2499 in 4 types
	Bitumen/rubber	Plasto-elastic	3–10	10–15	50	
	Pitch/polymer	Plasto-elastic	3–10	10–15	50	
Cold poured, 2 part chemically curing	Polysulphide	Elasto-plastic	5–20	15–25	50	Can be pitch modified
	Polyurethane	Elastic	5–20	15–25	50	
	Epoxy	Elasto-plastic	5–20	5–15	50	
Gun applied, non-curing	Oil based	Plastic	5–10	5–10	20	Forms surface skin
	Butyl	Plastic	5–15	5	20	Not recommended for exposed joints
	Acrylic	Plasto-elastic	10–20+	10–20	20	Available as solvent containing or aqueous emulsion
Gun applied, 1 part chemically curing	Polysulphide	Elasto-plastic	10–20	10–20	20	Primers required on porous substrates. Cure is initiated by atmospheric moisture. Cure rate slower than for 2 part materials
	Polyurethane	Elastic	10–20	15–30	20	
	Silicone	Elastic	10–20	15–30	20	
Gun applied, 2 part chemically curing	Polysulphide	Elasto-plastic	20+	20–30	25	BS 4254 applies. Primers required on porous substrates
	Polyurethane	Elastic	10–20	20–30	25	Primers required on porous substrates
	Polyepoxide/Polyurethane	Elasto-plastic	20+	50	50	
Hot applied, non-sag	Bitumen	Plastic	5–10	5–10	25	
	Bitumen/rubber	Elasto-plastic	5–10	10–15	25	
Strip	Butyl	Elasto-plastic	10–15	n/a	n/a	Suitable for situations where compression is provided in assembly
	Bitumen/rubber		10–15	n/a	n/a	
	Polyiso-butylene/butyl		15–20	n/a	n/a	
Hand applied, bedding compounds	Oil based	Plastic	5–10	Negligible	n/a	Will tend to dry on exposure to air. The life may be extended by painting. Generally unsuitable with aluminium windows

Notes

* Plastic materials exhibit plastic flow and have little or no recovery after deformation.
* Elastic materials (elastomeric) have physical properties similar to rubber and return to the original shape after deformation.
* Elasto-plastic and plasto-elastic exhibit partial elastic and partial plastic properties.

1 Movement accommodation is expressed as the total reciprocating movement occurring at the joint. Manufacturers may quote either this or ± figure based on the median joint width.
2 Multi-component sealants require mixing prior to application and must be used within their pot life.
3 Elastic and elasto-plastic sealants generally require primers on porous substrates.
4 This table is not an exclusive list of sealant materials available.

DESIGN FOR MOVEMENT IN BUILDINGS

Table 4.5 Suitability of various sealant types Based on PSA Advisory Leaflet no. 70, *Flexible ceiling materials* (no longer published)

Sealant application	Expansion joint (not subject to traffic)	Joints between cladding units	Pointing of window and door frames	Curtain wall joints	Glazing	Bedding window and door frames	Joints subject to traffic	Water retaining structures
Hot poured							●	●
Cold poured 2 part chemical curing							●	●
Hand applied, bedding compound					●	●		
Gun applied, non-curing		●	●	●	●			
Gun applied, 1 part chemical curing		●	●	●	●			
Gun applied, 2 part chemical curing	●	●	●	●	●			●
Hot applied, non-sag								
Strip					●	●		

Table 4.6 Width of sealant-filled joints

Total of contraction/ extension movement	Minimum width of joint for sealants which tolerate a total movement of		
	10%	20%	30%
mm	mm	mm	mm
2	16	8	7
3	32	16	10
5	48	24	15

Note Reference should be made to Table 4.5 to determine which sealants are capable of being used in the widths mentioned above. For example, 32 mm is too wide for some types of sealant since they would tend to sag.

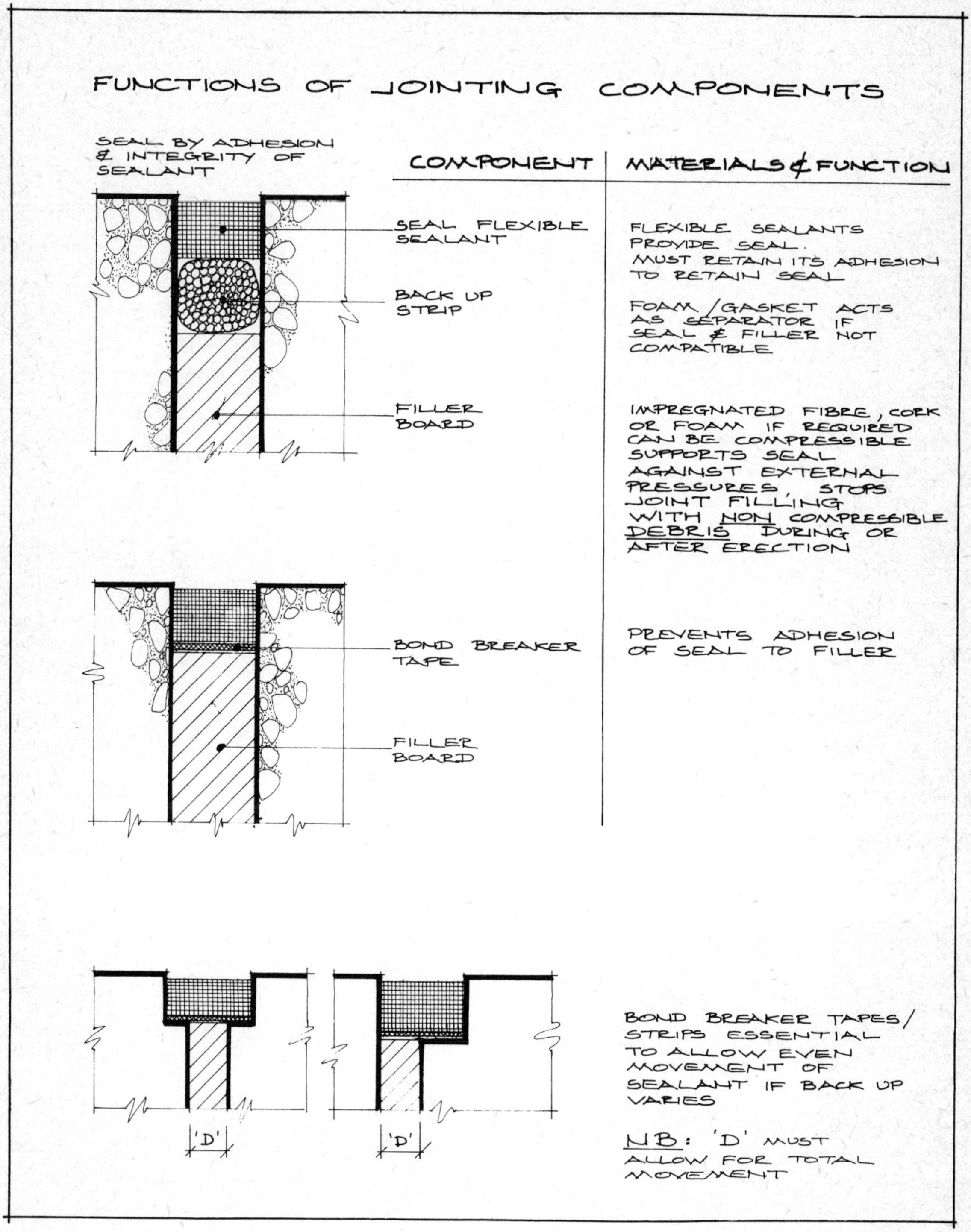

FUNCTIONS OF JOINTING COMPONENTS

SEAL BY ADHESION & INTEGRITY OF SEALANT

COMPONENT	MATERIALS & FUNCTION
SEAL FLEXIBLE SEALANT	FLEXIBLE SEALANTS PROVIDE SEAL. MUST RETAIN ITS ADHESION TO RETAIN SEAL
BACK UP STRIP	FOAM/GASKET ACTS AS SEPARATOR IF SEAL & FILLER NOT COMPATIBLE
FILLER BOARD	IMPREGNATED FIBRE, CORK OR FOAM IF REQUIRED CAN BE COMPRESSIBLE SUPPORTS SEAL AGAINST EXTERNAL PRESSURES, STOPS JOINT FILLING WITH <u>NON</u> COMPRESSIBLE <u>DEBRIS</u> DURING OR AFTER ERECTION
BOND BREAKER TAPE	PREVENTS ADHESION OF SEAL TO FILLER
FILLER BOARD	

BOND BREAKER TAPES/STRIPS ESSENTIAL TO ALLOW EVEN MOVEMENT OF SEALANT IF BACK UP VARIES

NB: 'D' MUST ALLOW FOR TOTAL MOVEMENT

Diagram 4.10

DESIGN FOR MOVEMENT IN BUILDINGS

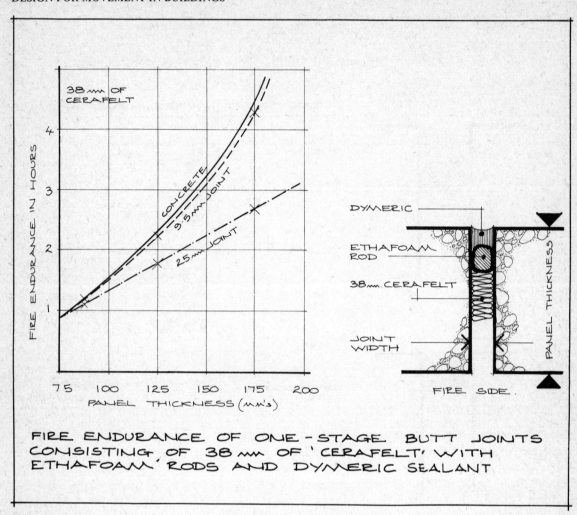

Diagram 4.11

MOVEMENT JOINT DESIGN

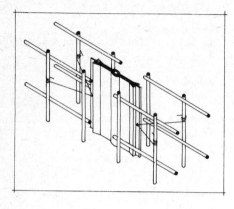

ISOMETRIC VIEW OF INTERNAL WATER STOP AT SHRINKAGE OF CONSTRUCTION JOINT IN VERTICAL R.C. WALL SHOWING DIFFICULTY OF KEEPING WATERBAR IN POSITION FOR CASTING

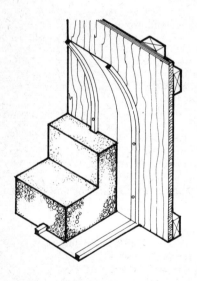

ISOMETRIC VIEW OF JUNCTION BETWEEN HORIZONTAL AND VERTICAL EXTERNAL (REAR) APPLIED WATERSTOP IN CONCRETE CONSTRUCTION SHOWING COMPARATIVE EASE OF POSITIONING WATERSTOP BY DIRECT FIXING TO SHUTTERING

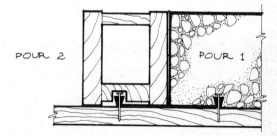

POUR 2   POUR 1

DIAGRAMS BASED ON MESSRS 'EXPANDITE' WATERSTOPS APPLICATION GUIDE

FORMATION OF JOINTS IN CONCRETE CONSTRUCTION

Diagram 4.12

Table 4.7 General list of joint fuctions, grouped under design aspects

Based on ISO 3447 and Table 2 BS 6093:1981

1 Environmental factors
1.1 To control passage of:
  (a) insects and vermin
  (b) plants, leaves roots, seeds and pollen
  (c) dust and inorganic particles
  (d) heat
  (e) sound
  (f) light
  (g) radiation
  (h) air and other gases
  (j) odours
  (k) water, snow and ice
  (l) water vapour
1.2 To control condensation.
1.3 To control generation of
  (a) sound
  (b) odours.

2 Capacity to withstand stress (either during or after assembly)
  To resist stress in one or more directions due to:
  (a) compression
  (b) tension
  (c) bending
  (d) shear
  (e) torsion
  (f) vibrations (or any other type of stress which may induce fatigue)
  (g) impact
  (h) abrasion (indicate, for each particular case, the type of wear)
  (j) shrinkage or expansion
  (k) creep
  (l) dilation or contraction due to temperature variations

3 Safety
3.1 To control passage of fire, smoke, gases, radiation and radioactive materials
3.2 To control sudden positive or negative pressures due to explosion or atmospheric factors
3.3 To avoid generation of toxic gases and fumes in case of fire
3.4 To avoid harbouring or proliferation of dangerous micro-organisms

4 Accommodation of dimensional deviations
4.1 To accommodate variations in the sizes of the joint at assembly due to deviations in the sizes and positions of the joined components (induced deviations)
4.2 To accommodate continuing changes in the sizes of the joint due to thermal, moisture and structural movement, vibration and creep (inherent deviations)

5 Fixing of components
5.1 To support joined components in one or more directions.
5.2 To resist differential deformation of joined components
5.3 To permit operation of movable components

6 Appearance
6.1 To have acceptable appearance
6.2 To avoid:
  (a) promotion of plant growth
  (b) discoloration due to biological, physical or chemical action
  (c) all or part of the internal structure showing
  (d) dust collection

7 Economics
7.1 To have a known first cost.
7.2 To have a known depreciation
7.3 To have known maintenance and/or replacement costs

8 Durability
8.1 To have specific minimum life, taking into account cyclic factors
8.2 To resist damage or unauthorised dismantling by man
8.3 To resist abrasive action
8.4 To resist action of:
  (a) animals and insects
  (b) plants and micro-organisms
  (c) water, water vapour or aqueous solutions or suspensions
  (d) polluted air
  (e) light
  (f) radiation (other than light)
  (g) freezing of water
  (h) extremes of temperatures
  (j) airborne or structure-born vibrations, shock waves or high-intensity sound
  (k) acids, alkalis, oils, fats and solvents

9 Maintenance
9.1 To permit partial or complete dismantling and reassembly
9.2 To permit replacement of decayed jointing products.

10 Ambient conditions
10.1 To perform desired functions over a specific range of:
  (a) temperature

(b) atmospheric humidity
(c) air or liquid pressure differential
(d) joint clearance variation
(e) driving rain volume

10.2 To exclude from the joint if performance would be impaired:
(a) insects
(b) plants
(c) micro-organisms
(d) water
(e) ice
(f) snow
(g) polluted air
(h) solid matter

Note Any one joint will have to satisfy a selection of functions only. This list cannot be comprehensive and the designer may have to identify additional functions applying in a specific situation.

5 Design for Movement Control in Building Elements

5.1 FOUNDATIONS

SCOPE

This section is intended to acquaint the reader with the significant factors affecting the design of foundations substructures to avoid settlement or movement. The actual selection and detailing of foundation structures is the province of the Structural Engineer who should carry out this work in all but the simplest strip footings.

The problem is very concisely stated in BRE Digest 63 namely:

'A building, its foundation and the supporting soil interact with one another in a complex manner, the behaviour of one depending upon and influencing, that of the others.'

Foundation design must therefore take into account not only the type of structure to be supported, its function and the constructional materials to be used, but also the soil conditions on site.

DESIGN CRITERIA AND FAILURE LIMITS

Building Regulation D 3, Foundations, states that the foundations of a building shall

(a) Safely sustain and transmit to the ground the combined dead load, imposed load and wind load *in such a manner as not to cause any settlement* or other movement which would impair the stability of or cause damage to the whole or any part of the building or of any adjoining building or works.
(b) be taken down to such a depth or be so constructed as to safeguard the building against damage by *swelling, shrinking,* or *freezing* of the subsoil; and
(c) be capable of adequately resisting any attack by sulphates or any other deleterious matter present in the subsoil. D 4 and D 5, 6 and 7 give deemed to satisfy provisions for reinforced concrete foundations and strip footings. The requirements of D 7 are illustrated in diagram 5.1.1.

Deformation limitations

In order to comply with this *general* requirement it is necessary to define *acceptable* limits for deformation, ie which will not 'impair the stability or cause damage'.

(a) Relative parts of the same building, ie differential vertical movement and deflections and
(b) overall angular distortion, ie tipping of the building at an angle to the vertical.

Deflection limits

In loadbearing masonry walls, the maximum deflection limits are determined by the ability of masonry to withstand tension and shear cracking. This is often determined by specifying the angular distortion. These limits are illustrated in diagram 5.1.2 and are expressed as the ratio between the deflection $\triangle$ and the length of wall L, ie $\frac{\triangle}{L}$.

Where walls only are affected, the limit can be as high as $\frac{1}{300}$, but in panel walls damage could result at a ratio of $\frac{1}{150}$, eg 20 mm over a 3 m span. Under $\frac{1}{150}$, damage to a supporting structural frame is unlikely.

Sagging deflection generally causes only small shear cracks between courses, whereas lagging deflection angles are usually half the sagging deflection angles (see diagram 5.1.2).

Classification of building damage

Reference must also be made to the BRE classification of building damage given below. It is unlikely that cracks with category 3 or higher damage would result if the angular distortion limits given above are adhered to.

Table 5.1.1 BRE Classification of building damage

Category of damage	Degree of damage	Description of typical damage* Ease of repair is shown in italics†	Approximate crack width mm†
		Hairline cracks of less than about 0.1 mm width are classed as negligible	≯ 0.1
1	Very slight	*Fine cracks which can easily be treated during normal decoration.* Perhaps isolated slight fracturing in building. Cracks rarely visible in external brickwork	≯ 1.0
2	Slight	*Cracks easily filled.* Redecoration probably required. Recurrent cracks can be masked by suitable linings. Cracks not necessarily visible externally; *some external repointing may be required to ensure weathertightness.* Doors and windows may stick slightly	≯ 5.0
3	Moderate	*Cracks require some opening up and can be patched by a mason. Repointing of external brickwork and possibly small amount of external brickwork to be replaced.* Doors and windows sticking. Service pipes may fracture. Weathertightness often impaired	5 to 15 or a number of cracks ⩾ 3.0
4	Severe	*Extensive repair work involving breaking-out and replacing sections of walls, especially over doors and windows.* Window and door frames distorted, floor sloping noticeably‡ Walls leaning‡ or bulging noticeably. Some loss of bearing in beams. Service pipes disrupted	15 to 25 but also depends on number of cracks
5	Very severe	*This requires a major repair job involving partial or complete rebuilding.* Beams lose bearings, walls lean badly and require shoring. Windows broken with distortion. Danger of instability	Usually > 25 but depends on number of cracks

* It must be emphasised that in assessing the degree of damage, account must be taken of the location in the building or structure where it occurs and also of the function of the building.
† Crack width is one factor in assessing degree of damage and should not be used on its own as a direct measure of it.
‡ Local deviation of slope, from the horizontal or vertical, of more than 1 in 100 will normally be clearly visible. Overall deviations in excess of 1 in 150 are undesirable.

SOURCES OF MOVEMENT: SOIL AND SUPERSTRUCTURE

Foundations transfer the loads of the structure and superimposed loads to the ground.

The characteristics of the soil are a variable factor between different sites and often on the same site and are the main source of movement. However the nature of the structure and fabric must also be taken into consideration both for its ability to withstand movements, sources of weakness and differential loading effects on an otherwise even subsoil.

When a combination of causes interacts to produce movement this becomes difficult to assess.

DESIGN FOR MOVEMENT CONTROL IN BUILDING ELEMENTS

REQUIREMENT OF BUILDING REGULATIONS D7 FOR UNREINFORCED FOUNDATIONS

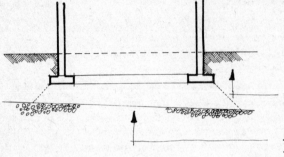

- NO MADE UP GROUND OR WIDE VARIATION IN TYPE OF SUBSOIL
- NO WEAK SOIL BELOW BEARING SOIL

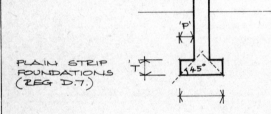

PLAIN STRIP FOUNDATIONS (REG D.7.)

'T' IS NOT TO BE < 'P' AND NOT LESS THAN 150 mm

CONCRETE TO B.S.882 PART 2 (50 Kg CEMENT + ⊅ 1m³ FINE & ⊅ 0.2 m³ COURSE AGGREGATE

WIDTH AS SPECIFIED IN TABLE TO BUILDING REGULATIONS

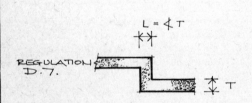

REGULATION D.7.

PLAIN UNREINFORCED STRIP FOUNDATION LAID AT MORE THAN ONE LEVEL

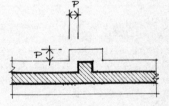

PROJECTION OF FOUNDATION ROUND PIERS SAME AS REST OF WALL

FOUNDATIONS : DETAILED DESIGN OF STRIP FOUNDATIONS TO ENSURE CORRECT DISTRIBUTION OF LOADS AND STRESSES

Diagram 5.1.1

FOUNDATIONS

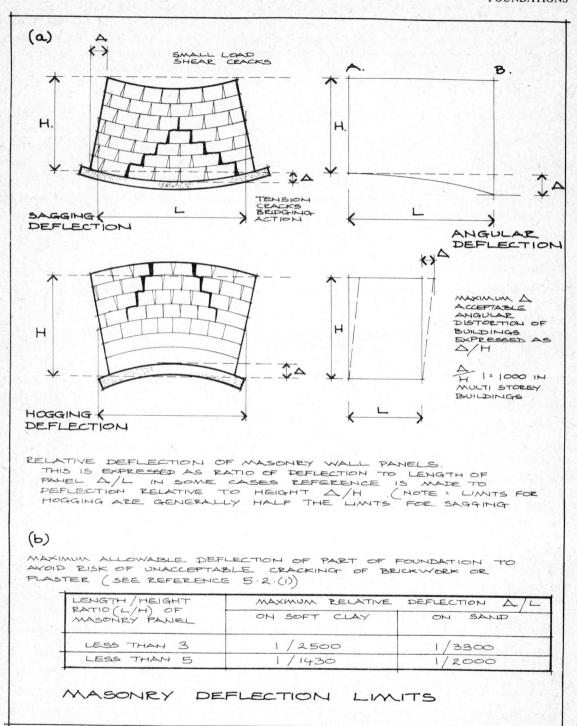

Diagram 5.1.2

Different factors may be significant in lightly loaded or heavily loaded structures on the same soil type. The actual movement will therefore depend on intrinsic soil characteristics but also on a combination of both loading and other extrinsic non-load related factors all of which will have to be taken into account in the selection and design of a foundation system.

Characteristics of soils

The movement characteristics of different soils vary greatly both as regards the amount of movement and the duration in time taken for the movement.

Intrinsic movements: shrinkage and swelling

Shrinkage and swelling of soils usually occurs due to the raising or lowering of the water content. The amount of water in proportion to solid matter varies greatly. It is high in clays and low in sands consequently swelling is high in clays and low in sands. Seasonal changes due to the presence of vegetation only extends some 1.5 m deep with significant shrinkage only in the top 1 m. However the desiccation caused by trees can extend far lower than this.

Extrinsic movements: consolidation and swelling

Consolidation occurs when the soil is subjected to extra pressures either due to foundations or extra soil. Decompression can result in swelling. The basic mechanism of consolidation is as follows: the water is squeezed out by pressure and soil particles are forced together.

This occurs quickly in sands and slowly over a long period in clays, which can cause problems during the construction process.

Combined effects

Where soil is removed and replaced with building loads a combination of swelling and consolidation can occur.

Shrinkable clay soils

Clay is defined as soil containing a large proportion of small particles under 0.002 mm, giving them the characteristic smooth texture. Wet clays are soft and sticky. Dry clays are very hard and brittle. Clays vary in their acceptance of water by reason of mineral type as well as particle size.

Tests to BS 1377

These tests define the clay nature and plasticity.

Liquid limit test measures the percentage of water to dry weight when the addition of water has caused the sample to reach a liquid state.

Plastic limit test measures the amount of water as a percentage, when the clay ceases to be plastic, ie breaks up.

Prediction of shrinkage and swelling of soils

A indirect method, ie other than direct measurements on site, is to examine, sample and establish its *potential* to shrink and swell.

The tests to determine this are given in BS 1377; they should be carried out by a soil testing laboratory.

The tests establish

1 *Plasticity index*
 (subtracting plastic limit from liquid limit)
2 *Clay fraction*
 percentage of soil with particle size less than 0.002 mm.

Shrinkage potential

Table 5.1.3, taken from BRE Digest 240 gives the shrinkage potential of various clays:

Table 5.1.2 Clay shrinkage potential

Plasticity Index	Clay fraction	Shrinkage potential
>35	>95	Very high
22–48	60–95	High
12–32	30–60	Medium
<18	<30	Low

FOUNDATIONS

SUMMARY OF CAUSES OF NON-LOAD RELATED MOVEMENTS

1 *Extrinsic causes* (see diagrams 5.1.3)

Variation in water content by:
Planting of trees
Removal of trees
Proximity of trees } diagram 5.1.3
Seasonal changes
Consolidation due to changes of water content or variations in ground water
Due to flooding or
other causes such as new load surfaces subsoil and new drainage.

Changes in compaction and support by:

Excavations; removal of soil
Mining operations
Disturbance from adjoining sites or works
Sliding action of clay on a slope or natural erosion, landslip.

2 *Intrinsic causes*

Weak soil or variable strata deep fill or imported filling; uneven support, eg over old foundations or basements; high ground water level, eg marshy ground.
Changes of soil moisture content due to trees, changes of water table in exceptionally dry seasons.

CAUSES OF LOAD RELATED MOVEMENTS
(see diagrams 5.1.4 and 5.1.5)

Settlement due to the imposition of loading.
Differential settlements due to uneven loading or uneven bearing capacity of the soil or weaker layers within the stress envelope (see diagram 5.1.5 for the extent of bulbs of pressure under various types of foundation).

MOVEMENT CONTROL METHODS

Treatment of soils and fills

The shortage of building land with good firm natural and uniform supporting ground has meant that more development is taking place on ground or fill of variable or poor loadbearing capacity. Alternatively the loadbearing capacity of firmer ground can be improved by treatment to take greater loads without undue settlement. This allows the use of simple strip footings instead of piles, which would otherwise be necessary. Although high-rise buildings may still require piling, suitable fill after treatment can support five to six storey structures on conventional footings imposing a bearing pressure of the order of 160 kn/m². Hydraulic sand fill can be compacted to give bearing pressures up to 400 kn/m².

Table 5.1.3 Shrinkage potential of some common clays

Clay type	Plasticity Index %	Clay fraction %	Shrinkage *potential*
London	28	65	Medium/high
London	52	60	High
Weald	43	62	High
Kimmeridge	53	67	High/very High
Boulder	32	—	Medium
Oxford	41	56	High
Reading	72	—	Very high
Gault	60	59	Very high
Gault	68	69	Very high
Lower Lias	31	—	Medium
Clay silt	11	19	Low

Types of natural soils requiring treatment

These are types of soil which would not allow the use of ordinary spread footings. These fall into two main categories: cohesive and non-cohesive. The cohesive soils are usually alluvial clay deposits. They are difficult to compact, and treatment usually consists of the introduction of stone 'columns' on a fairly regular grid under the proposed foundations and usually taken down to a firmer sub-base or more compacted layers.
Non-cohesive or granular soils such as naturally occurring deposits of sand can vary greatly in density over the site, especially if they include previous excavated material or fill. Treatment has to ensure that consolidation is achieved in such a manner over the whole site as to give a consistant bearing pressure, that is, forming a 'raft' of

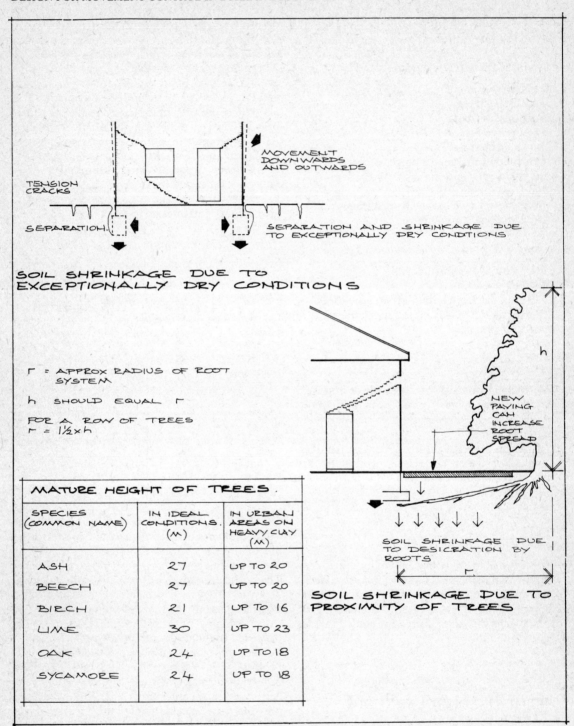

Diagram 5.3

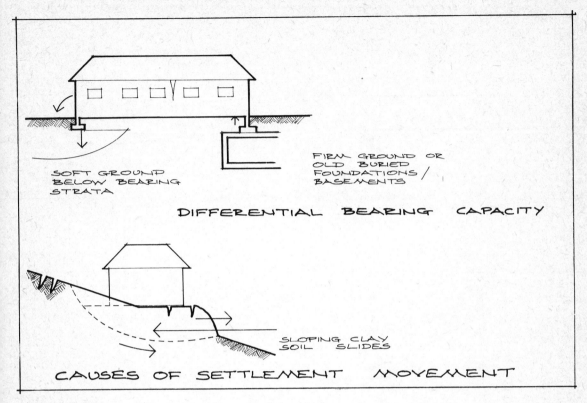

Diagram 5.1.4

compacted ground of uniform and high density under the proposed structure to distribute the loads from spread footings in the normal manner and to be deep enough to bridge over any possible pockets of variable density at depth.

Types of fill

All types of fill are subject to considerable settlement movement even under self weight and this movement can take place over a period of five to seven years and, in some cases, up to 20 years. Not only the type of fill but its age must be taken into consideration.

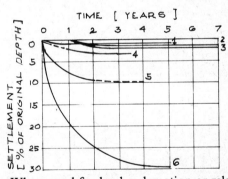

Curve	State of compaction	Material
1	Well compacted	Well graded sand
2	Medium	Rock
3	Lightly compacted	Clay
4	Uncompacted	Sand
5	Uncompacted	Clay
6	Well-compacted	Mixed refuse

Where used for land reclamation or relevelling the method of depositing has probably been carefully controlled. This may not be the case in random filled sites such as over demolished buildings, etc. The settlement potential varies greatly with the type of fill and method of placing. The permeability, rate of consolidation and organic content, as well as watertable affect the rate of settlement and compressibility of the fill making predictions difficult.

DESIGN FOR MOVEMENT CONTROL IN BUILDING ELEMENTS

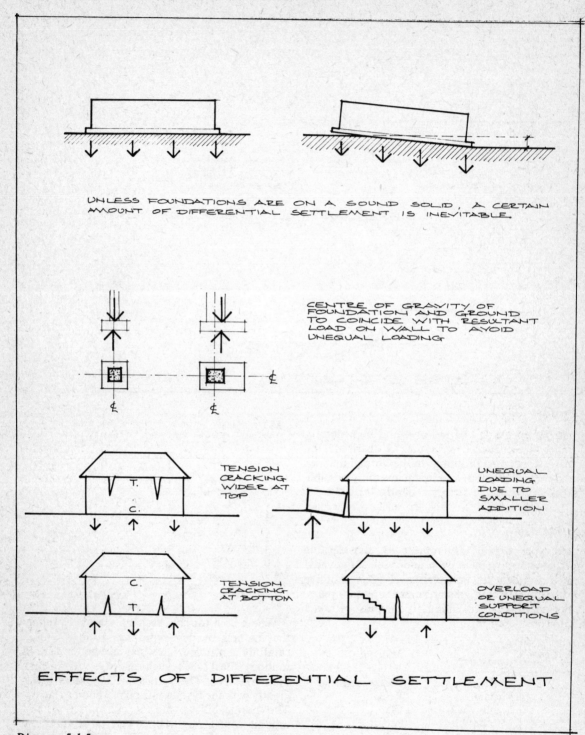

Diagram 5.1.5

It is essential to dig trial pits and if necessary carry out compression tests both before and after treatment to find the bearing capacity and likely settlement.*

Types of fill:

1. Natural granular fill – sand, gravel, crushed rock shale
2. Natural cohesive fill – clay and clayey silts
3. Industrial waste – slag ash, mining waste, demolition tips, chemical tips
4. Domestic refuse – containing various proportions of organic substances

Treatment methods

Cohesive soils are usually modified by partial replacement with stronger materials introduced by mechanical means. Non-cohesive or granular soils are usually deeply compacted. This is achieved by the following methods:

1. *Non-specialist techniques*
   Pre-compression: Pre-loading in excess of the final loading. This requires a time of up to 100 days or over
   Ground water lowering: Mainly used for granular fills where the water table is high or to increase the density of loose granular material
   Surface rolling or ramming: Mainly for granular fill of not over 1 m thickness *maximum*
2. *Specialist techniques*
   Grouting voids: Pressure grouting of fissures and voids can be useful if vibratory methods are unacceptable due to surrounding properties and to fill voids in loose or uneven fills
   Explosive compaction: Mainly used in USA and USSR in very open sites or under the seabed
   Dynamic consolidation: This process uses a high energy compaction or surface tamping. Ground of up to 6 m can be consolidated with weights of up to 20 tonnes dropped from heights of 20 metres. New fill has been treated at Surrey Docks by this method to provide sites for two-storey housing, with conventional strip footings.

* For a fuller treatment of the subject the reader is referred to the articles given in the Reference section by specialists in this field.

Vibro-flotation: This is a technique for vibro-compaction or vibro-replacement of parts of the soil below the surface by the use of a vibro-float tube suspended by a crane with an eccentric moving head to consolidate the ground. It may, or may not, have a 'jetting' fluid to aid both penetration and consolidation. Columns of granular material can also be introduced into the ground at regular centres. These columns also act as drains increasing the natural ground strength. The spacing of the compaction centres to give overlapping or 'bridging' zones to cover a wide area depends on ground conditions and loads and is determined by the specialist. When used in cohesive soils these 'stone columns' properly spaced can reduce the compressibility of the soil by 40% and greatly increase the shear resistance of the soil.

When properly carried out by specialist firms these treatments have proved adequate on treated refuse to allow bearing pressures of up to 200 kn/m^2 and flats of up to 6 storeys have been built on treated granular building debris. It is claimed that in many cases a saving of 25% to 50% on piling can be achieved.

SELECTION OF FOUNDATION TYPES (IN SHRINKABLE CLAY SOILS)

Building location: Open ground

without danger of deep rooted vegetation causing movement deeper than 0.9 m.

Strip or narrow strip (trench fill)

normal strip at 0.9 m or narrow deep strip, unless there is a possibility of *future* deep rooted vegetation when short bored piles can be used say 4.5 m deep depending on soil conditions and loading

Near major existing vegetation

The risks are greater near roots or groups of trees.

DESIGN FOR MOVEMENT CONTROL IN BUILDING ELEMENTS

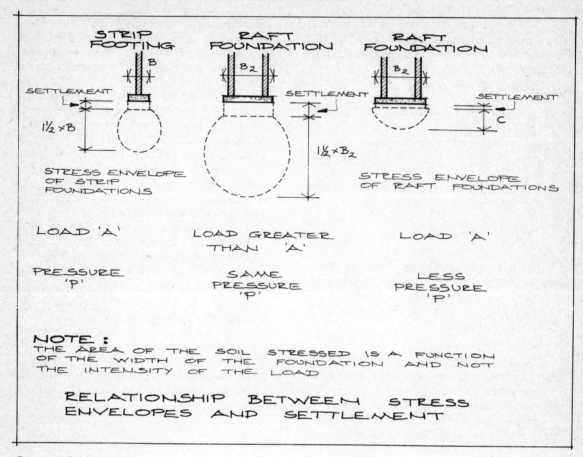

Diagram 5.1.6

Root spread should be checked by trial pits.

Bored piles are necessary to depths below desiccation level. This can be 5 in. or more. The top section of the piles may require sleeving to avoid uplift by swelling of the soil (see diagram 5.1.8)

Where trees have been removed

Swelling can go on for years after trees are removed.

Piles as for existing trees are necessary. In addition, the upper part of the piles must be sleeved (see diagram) to avoid uplift due to swelling, or the pile lengthened to compensate for uplift by extra friction.

The ground beams must be cast on a layer of compressible material.

The ground floor slabs should also be suspended over a layer of compressible material or a gap created by a self-degrading layer of fibreboard or specialist material, eg 'clayboard'.

5.2 BASEMENTS AND SUBSTRUCTURES: and elements in contact with the ground

SCOPE

This section covers elements of construction at ground level and where in contact with the ground or on fill, hardcore and basements.

Buildings on fill

Where the entire building, ie the foundations are constructed on hardcore or fill the problem of both

BASEMENT AND SUBSTRUCTURES

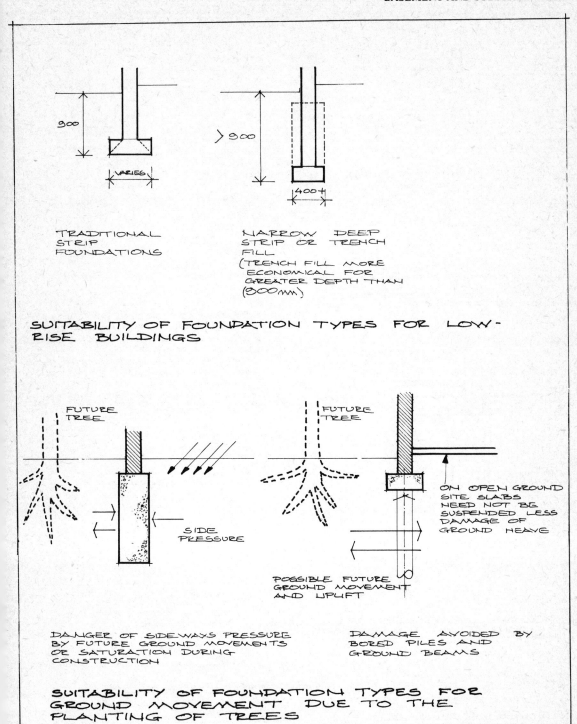

Diagram 5.1.7

DESIGN FOR MOVEMENT CONTROL IN BUILDING ELEMENTS

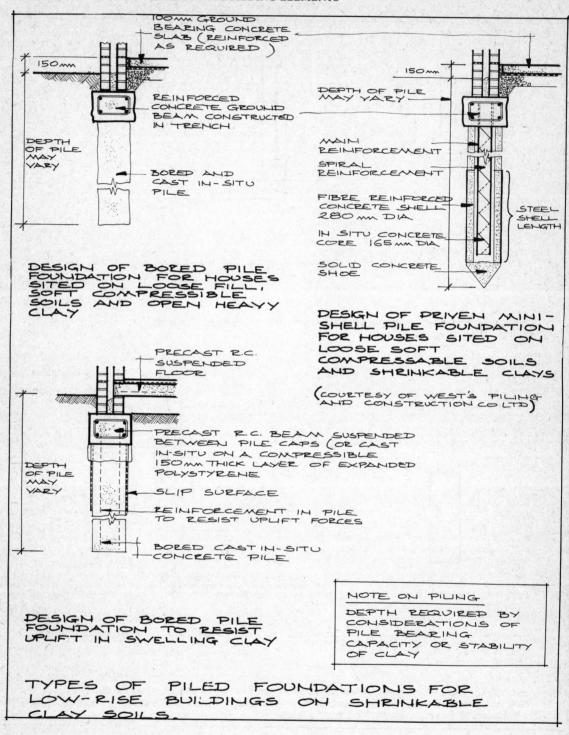

Diagram 5.1.8

BASEMENT AND SUBSTRUCTURES

initial and progressive movement of foundations can be serious and requires expert advice. In such cases structural and foundation systems vary widely from rafts to large area concrete 'pad' formations for light framed buildings and deep piling and ground beams for heavy masonry structures.

Differential movements between the lowest floor and the rest of the structure are likely to occur and require special attention.

GROUND FLOOR SLABS: DESIGN CRITERIA

Classification of effects and damage

As with other elements it is necessary to establish the limits of acceptable damage. The following table from BRE Digest 251 'Assessment of damage in *low rise buildings*' classified damage effects due to ground floor slab failures. These are due mainly to movement in the supporting fill. The classifications are based on the level of the costs of repairs. Categories 0–2 could well be due to poor compaction and may not be progressive. Categories 3–5, being due possibly to actual voids below the slab or chemical action in the fill, could be progressive. See table 5.2.1.

Incidence of failure of ground floor slabs

BRE Digest 176 *Failure patterns and implications* gives a table analysing 256 defects other than damp which shows the high proportion of movement 'cracking' failure in ground floors as compared with others.

Type of defect	Total no.	Location	
Cracking	93	due to foundation movement	8
		in floors:	28
		in loadbearing walls:	20
		in non loadbearing walls:	10
		in wall and floor finishes:	14
		other	13
		Total	93

Serviceability failures

Movement failures in ground floor slabs are therefore usually detected by relative movements with the enclosing walls, rather than cracks in the slab itself showing more serious support failures. Detachment and failure of 'hard' finishes or tiles being an associated failure.

However if movement is slight and not progressive this may not be serious.

Since ground floors are not required to act as water retaining structures distributed fine cracking should not lead to problems of loss of damp proofing as the damp proof membrane usually is flexible enough to cover these.

Crack toleration of screeds

In fact, a series of fine hair cracks in the top of the slab can usually be tolerated by normal cement/sound screeds. This is the limitation for which normal slab construction and reinforcement is designed.

GROUND FLOOR SLABS: CAUSES OF MOVEMENT

Foundation movement

Ground floor slabs and the associated hardcore under them can be affected by foundation movement and also in turn affect the foundations. However this depends on the construction of the ground floor slab. If suspended or tied into a ground beam, any foundation movement will seriously affect the ground floor slab (see diagram 5.2.1).

Other causes: load-related deflections and failures

Load related movements: diagram 5.2.2 shows the loading effects of internal partitions, which could cause failure due to fracture or excessive deflection in the slab.

Differential settlement

Inequalities in the hardcore or fill can lead to differential movement/settlement of the slab. By far the most serious problems in ground floor slabs are those where slabs are partially constructed on fill and partially on existing ground.

If a clear definition can be made between each section at a dividing wall this can be avoided. However it is still safer to treat the entire area of

Table 5.2.1 Classification of visible damage caused by ground floor slab settlement

The classification below attempts to quantify the assessment of floor slab settlement damage in a similar way to that for superstructure damage, given in table 5.1.1. It has not yet been used extensively to determine its applicability. It should be noted that the categorisation may be qualified by the possibility of progression to a higher category; this should arise only when examination has revealed the presence of voids or areas of loosely compacted fill (or degradable material) beneath the floor slab such that more settlement can be expected.

Category of damage	Degree of damage	Description of typical damage	Approximate (a) crack width (b) 'gap'[1] mm
0	Negligible	Hairline cracks between floor and skirtings	(a) NA (b) up to 1
1	Very slight	Settlement of the floor, slab, either at a corner or along a short wall, or possibly uniformly, such that a gap opens up below skirting boards which can be masked by resetting skirting boards. No cracks in walls. No cracks in floor slab, although there may be negligible cracks in floor screed and finish. Slab reasonably level.	(a) NA (b) up to 6
2	Slight	Larger gaps below skirting boards, some obvious but limited local settlement leading to slight slope of floor slab; gaps can be masked by resetting skirting boards and some local rescreeding may be necessary. Fine cracks appear in internal partition walls which need some redecoration; slight distortion in door frames so some 'jamming' may occur necessitating adjustment of doors. No cracks in floor slab although there may be very slight cracks in floor screed and finish. Slab reasonably level.	(a) up to 1 (b) up to 13
3	Moderate	Significant gaps below skirting boards with areas of floor, especially at corners or ends, where local settlements may have caused slight cracking of floor slab. Sloping of floor in these areas is clearly visible. (Slope approximately 1 in 150) Some disruption to drain, plumbing or heating pipes may occur. Damage to internal walls is more widespread with some crack filling or replastering of partitions being necessary. Doors may have to be refitted. Inspection reveals some voids below slab with poor or loosely compact fill.	(a) up to 5 (b) up to 19
4	Severe	Large, localised gaps below skirting boards: possibly some cracks in floor slab with sharp fall to edge of slab; (slope approximately 1 in 100 or more). Inspection reveals voids exceeding 50 mm below slab and/or poor or loose fill likely to settle further. Local breaking-out, part refilling and relaying of floor slab or grouting of fill may be necessary; damage to	(a) 5 to 15 but may also depend on number of cracks (b) up to 25

continued

BASEMENT AND SUBSTRUCTURES

continued

Category of damage	Degree of damage	Description of typical damage	Approximate (a) crack width (b) 'gap'[1] mm
		internal partitions may require replacement of some bricks or blocks or relining of stud partitions.	
5	Very severe	Either very large, overall floor settlement with large movement of walls and damage at junctions extending up into 1st floor area, with possible damage to exterior walls, or large differential settlements across floor slab. Voids exceeding 75 mm below slab and/or very poor or very loose fill likely to settle further. Risk of instability. Most or all of floor slab requires breaking out and relaying or grouting of fill; internal partitions need replacement.	(a) usually greater than 15 but depends on number of cracks (b) greater than 25

Note[1] 'Gap' refers to the space – usually between the skirting and finished floor – caused by settlement after making appropriate allowance for discrepancy in building, shrinkage, normal bedding down and the like.

slabs as suspended in such cases (see diagram 5.2.3).

Damp-proof courses

The link between vertical damp-proof courses and horizontal damp-proof courses and damp-proof membranes can fracture and lead to damp penetration.

Chemical action on hardcore and fill

Expansion of the hardcore due to the action of water on the impurities contained in the hardcore can lead to serious swelling/expansion in the base hardcore with hogging failure in the slab. The problem is most serious where there is a high water table, or poor surface water drainage. It may be necessary to place a separating or blinding layer of clinker, blinded before laying the main hardcore.

Sulphates in hardcore

Sulphates in the hardcore should not exceed 0.5 per cent if used on a wet site, and in contact with ordinary Portland Cement concrete to avoid sulphate attack on the concrete with subsequent swelling and other deterioration effects on all concrete in contact with its ground floor slab and foundations.

Amount and duration of movement by fill/hardcore

The duration of settlement of fills/hardcore under load:

It should be emphasised here that the duration of the bulk of the movement is usually two years after laying of fill (see BRE Digest 142)

Diagrams 5.2.4A shows the layers recommended for non-structural ground supported concrete slab construction.

As settlement movement in even well compacted hardcore is difficult to predict and likely to be irregular, the NHBC practice No. 6 forbids the use of solid ground floors on hardcore over 600 mm in depth.

Intrinsic failures

These are rare in slabs under 6–7 m in any direction, ie most domestic buildings.

However for larger and more heavily loaded slabs, constructed in concrete, all the problems described in the section on materials/concrete will need attention as described in the following sections.

Movements in large RC supported slab floors

As distinct from unsupported, ie suspended slabs, these slabs are often the responsibility of the architect without the benefit of an engineer's

DESIGN FOR MOVEMENT CONTROL IN BUILDING ELEMENTS

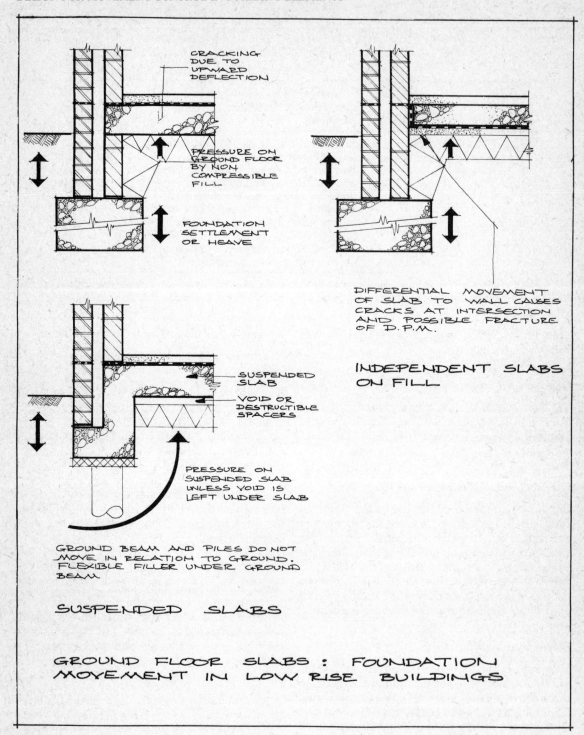

Diagram 5.2.1

BASEMENT AND SUBSTRUCTURES

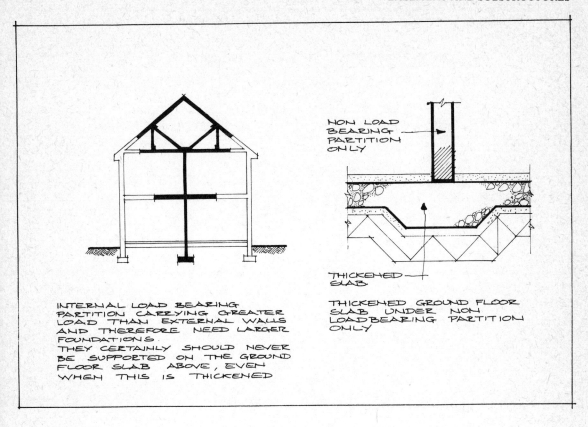

Diagram 5.2.2

advice. Although as stated earlier these slabs need not usually be constructed to watertight concrete standards. Nevertheless cracks could cause serviceability failure in finishes and possibly some damp penetration or warping deformations which are more difficult to repair.

CONCRETE GROUND FLOOR SLABS: SOURCES OF MOVEMENT

(a) Concrete early age shrinkage.
    Concrete early age thermal movement cracking.
(b) Differential settlement of hardcore base.
(c) Loading related failures generally.
(d) Restraint at columns or enclosing walls and differential settlement with adjoining elements at junctions.

SMALL GROUND SLABS ($\ngtr 7\,m \times 7\,m$): MOVEMENT CONTROL

In most domestic scale ground floor or lowest floor slabs, where these are fully supported on the ground the only necessary precaution against movement failure is in the quality of the sub-base and if applicable the subgrade.

Fill used for sub-grade should be as carefully selected as for the hardcore sub-base and well compacted.

The sub-base should consist of inert graded granular material of maximum size not exceeding 75 m (Building rubble needs breaking up!) fully compacted and blinded with sand or crushed fine material to form a smooth level surface with a tolerance of $+0-25\,m$ or be in lean mix concrete (wet site conditions or heavy construction traffic).

83

DESIGN FOR MOVEMENT CONTROL IN BUILDING ELEMENTS

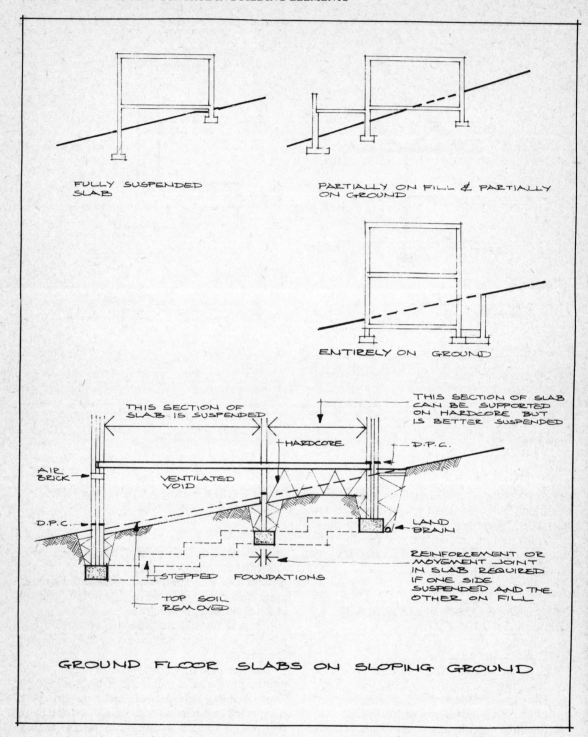

Diagram 5.2.3

BASEMENT AND SUBSTRUCTURES

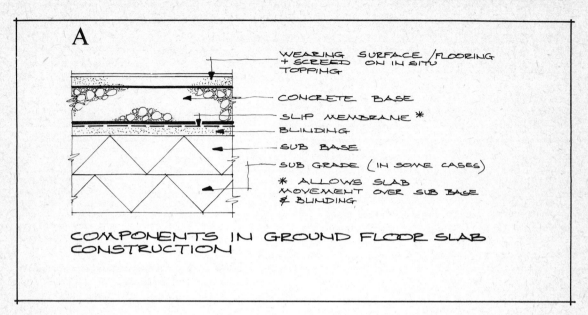

Diagram 5.2.4A

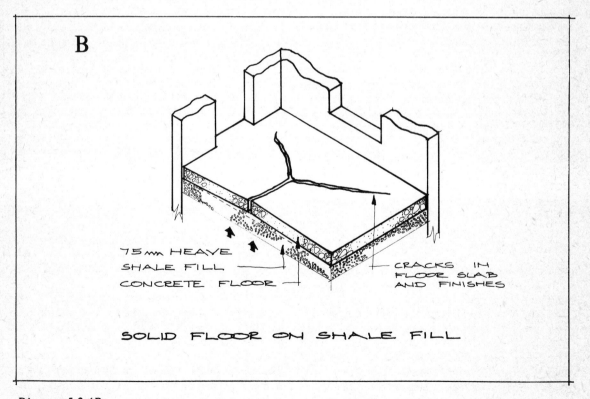

Diagram 5.2.4B

DESIGN AND CONSTRUCTION METHODS FOR MOVEMENT CONTROL IN LARGE AREA NON-STRUCTURAL CONCRETE SLABS ON THE GROUND

It is possible to avoid failures due to the intrinsic and extrinsic movements described earlier in this section by adopting the recommended precautions in design and construction. These have been clearly described by R Colin Deacon in the C and CA booklet *Concrete ground floors*, and the following is a summary of the recommendations. These are based on extensive and long experience in the construction of large concrete slabs for roads, for which they were originally developed and are now used extensively for reservoirs and large area and industrial floor construction.

'Long strip' method or bay method

Large floor slabs over about 5–7 m in any one direction have to be constructed in several separate jointed sections because of:

(a) practicability of reach for compaction
(b) control of tensile stresses due to thermal and moisture (shrinkage) movements.

Bay method The alternate bay method (see diagram 5.2.5) is not considered advisable, using mesh reinforcement through split formwork in the sides of the bays, because full access is difficult and joints are too numerous and difficult to form.

'Long strip' method As shown in illustration diagram 5.2.6 the method consists in laying the slab in alternative long strips, about 4.5–5 m in width with longitudinal construction joints between strips.

Crack control

Cracking due to early thermal effects and drying shrinkage are avoided by the provision of joints and reinforcement.

As the purpose of the reinforcement is *to control* (*not prevent*) *cracking* it is placed in the top 100 mm of the slab. A suitable weight of mesh should be used as recommended by table 5.2.2 from *Concrete Ground Floors* by C Deacon.

Reinforcement: Where shrinkage cracking has to be prevented to avoid serviceability failures.

Table 5.2.2 Fabric reinforcement for slabs of various thickness

Thickness of slab	BS standard mesh fabric to be used for maximum effective length of slab between free joints of				
(mm)	15 m	30 m	45 m	60 m	75 m
125	A 142	C 283		C 385	
150	A 142	C 283		C 385	C 503
175	A 142	C 283	C 385	C 503	C 636
200	A 142	C 283	C 385	C 636	
225	C 283	C 385	C 503	C 636	C 783

Notes
1 A 142 is a standard square-mesh fabric. The others are standard long-mesh fabrics.
2 If square-mesh fabrics only are obtainable, the cross-sectional area of main wires should be equivalent to that of the required long-mesh fabric.
3 If dowelled contraction joints are used in place of tied transverse joints, the required mesh weight is determined from the joint centres.

Example: A slab 200 mm thick has an effective length of 45 m between free isolation joints at its perimeter, with tied control joints at 10 m centres.

From this table, standard long-mesh C 385 is required. If dowelled joints at 15 m centres are substituted for the tied joints, the effective slab length becomes 15 m, and standard square mesh A 142 is suitable.

The table is based on using a single layer of mesh in the top surface zone equal to half the slab thickness not less than 100 m.

Jointing

Longitudinal joints (between strips) These are basically construction joints, but being only 4.5 m apart are usually constructed as 'tied contraction joints' (see diagram 5.2.6).

As it is not practicable to carry the reinforcing mesh across, tie bars are used to provide continuity and with heavily loaded slabs, double bars to prevent relative deflection movements.

Transverse joints These are formed across the strips. The spacing should be 6 m in unreinforced slabs and can be 10 m in reinforced slabs if fine

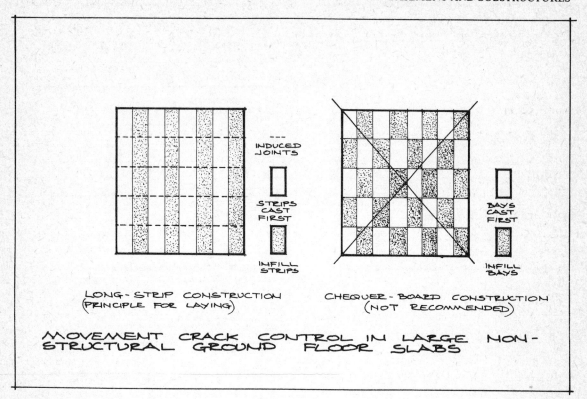

Diagram 5.2.5

cracking requires control. The tied transverse control joint is suitable for these joints as shown in diagram 5.2.6. It is basically a system for inducing a shrinkage crack on the line of the joint, but with continuity of reinforcement.

Expansion joints (see diagram 5.2.7) Using this system, expansion joints are only required every 70 m, so that the layout shown in diagram 5.2.6 can usually be used for the entire length of a building.

Other measures for crack control

Isolation joints Against perimeter walls, stancheon and plant bases shown in diagram 5.2.7.

1 subgrade and sub base construction to avoid settlement failures were described earlier.
2 Adequate thickness to avoid structural failure and excessive deflection, depends also on the sub grade quality, as shown in table 5.2.3 C and CA, *Concrete ground floors*, for general guidance.

Table 5.2.3

	Subgrade Classification	Recommended Thickness
Offices shops classrooms private garages	weak	175
light industrial 5KN/m²	normal	150
Commercial garages	weak	200
Industrial and warehouses 5–20 KN/m²	normal	175

Further guidance is given in this publication on slab thickness for fork lift truck wheeled traffic.

3 *Mix design, compaction, and curing*
The usual precautions in concrete construction against premature drying out, evaporation of bleeding water, frost attack, etc, must be taken as well as careful mix design and supervision.

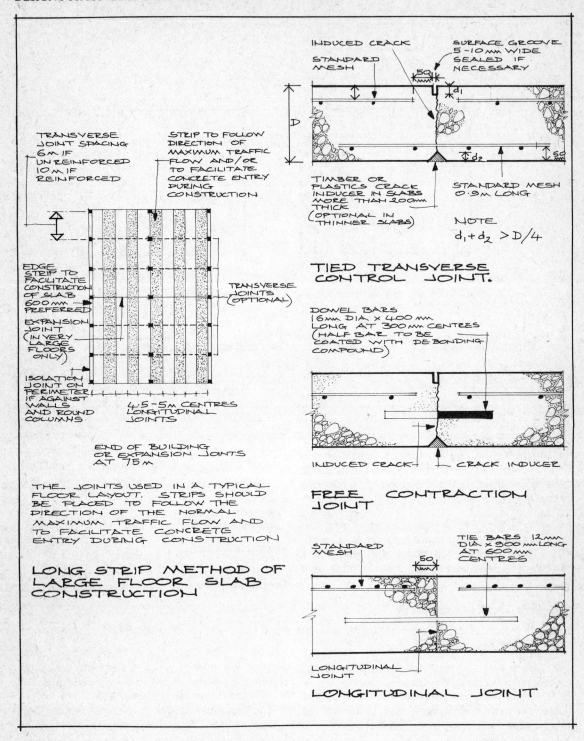

Diagram 5.2.6

BASEMENT AND SUBSTRUCTURES

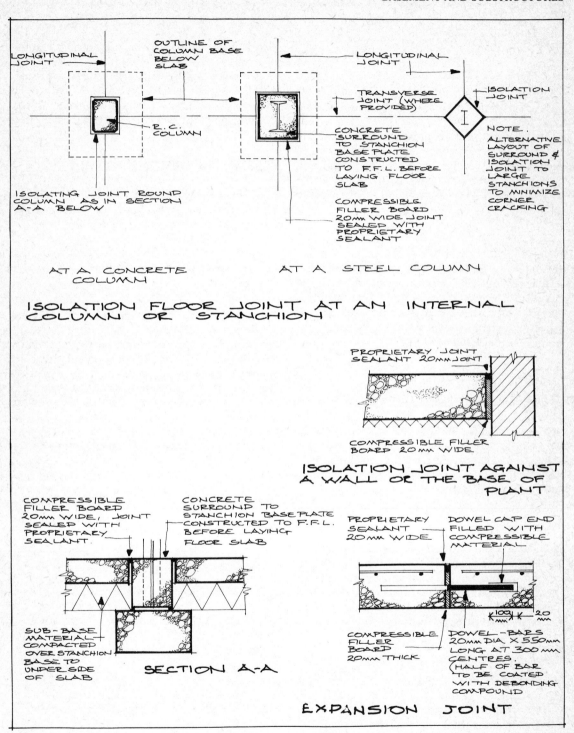

Diagram 5.2.7

RETAINING WALLS AND FREE STANDING WALLS

Types

These can be free standing, or forming part of an external unenclosed area, or part of an enclosed basement.

As the functional requirements and therefore the criteria for movement control are different in each of these situations, their section will deal with both the requirements failure patterns and design precautions against movement failure of each type separately.

Free standing walls: design criteria

Serviceability failures in free standing retaining walls are: loss of surface appearance and the integrity of applied facings and claddings. The tolerance of finishes to cracking will be the same as given for external walls. Naturally stability remains the most serious failure.

Surface and integrity of finish

In addition damp penetration from the ground behind could cause unsightly staining on the exposed surface so that in some cases these walls may require to be water resisting. However this would generally not be critical as in the case of walls forming part of a basement enclosure which will be dealt with later.

Causes of failure

Failures due to movement can be caused by movement in the ground or by movements due to excessive stresses in the wall construction, with a variety of causes. Using a previously described classification, failure due to movement can be caused by intrinsic or extrinsic factors.

Extrinsic causes of failure: the ground

Movement in the ground can cause the most serious type of failure, possibly structural and various types of failure of the ground and resulting movements imposed on the wall are illustrated in diagram 5.2.8 and a condition below which this is not likely that structural failure could occur.

Intrinsic causes: wall construction and material

Movements in the wall construction itself generally causes serviceability failures, unless water exclusion is critical, by leakage through fine cracks with subsequent surface staining. As with other reinforced concrete structures corrosion of the reinforcement can lead to spalling of the concrete.

BASEMENTS: DESIGN CRITERIA

Basements, in whole or part below ground, are usually required to be both soil retaining and water tight structures.

MOVEMENT FAILURE LIMITS: STRUCTURAL

1 Cracking due to excessive loading is rare since the design of the retaining structures is usually carried out in full compliance with the stability requirements of the building regulations by qualified Engineers. However it is useful to appreciate that basement walls and floors acting in this way are subject to fluctuating pressure distribution so that slabs and walls or bases all require careful integration to avoid cracking, with care in design and execution (see diagram 5.2.9).

Water retention

However the most likely cause of failure would be movement cracking to allow water penetration from the ground.

CP102:1973 type A and B structures

CP102:1973 *The protection of buildings against water from the ground* describes two basic options for design: Type A: Structures – requiring an impervious membrane. Type B: Structures without membranes.

BASEMENT CONSTRUCTION: MOVEMENT CONTROL METHODS

Whether a type A or B structure the intention should be to construct as monolithic and simple a

BASEMENT AND SUBSTRUCTURES

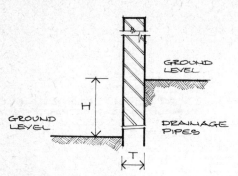

FREESTANDING WALLS IN LOAD BEARING MASONRY.

A RULE OF THUMB FOR PERMISSABLE CHANGE OF GROUND LEVEL IS H NOT > T × 3

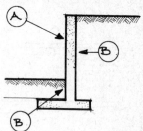

A SEVERE EXPOSURE
B MODERATE EXPOSURE

APPLICATION OF B.S. 5337 CLASSIFICATION OF EXPOSURE CONDITIONS TO A FREESTANDING RETAINING WALL IN REINFORCED CONCRETE

MIX DESIGN MUST ALLOW FOR APPROPRIATE CONDITIONS OF EXPOSURE TO AVOID EXCESSIVE SHRINKAGE CRACKS

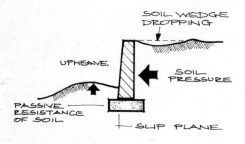

A FAILURE BY SLIDING FORWARD

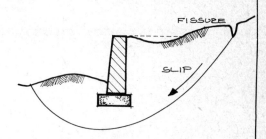

B FAILURE BY SOIL SLIP PLANE
(COHESIVE SOIL ONLY)

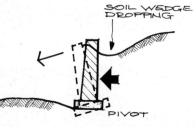

EXCESSIVE GROUND PRESSURE DUE TO EXCENTRIC LOADING

C FAILURE BY OVERTURNING

RETAINING WALLS : FAILURE DUE TO SOIL MOVEMENT AND EXPOSURE

Diagram 5.2.8

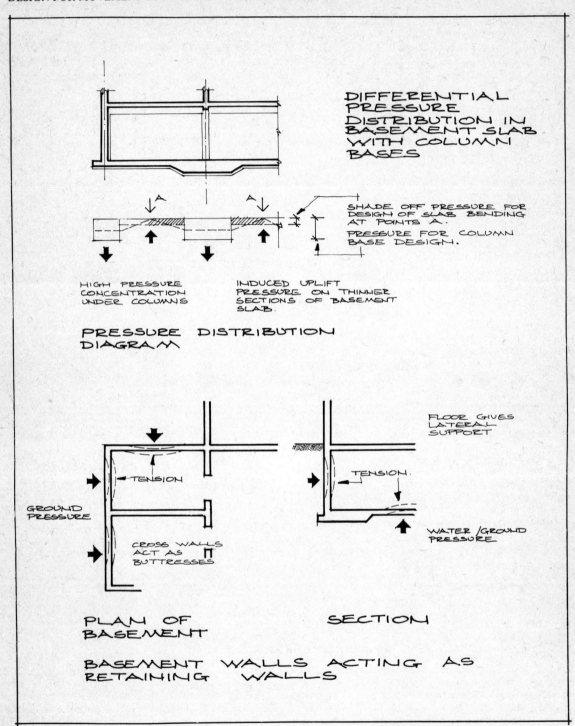

Diagram 5.2.9

BASEMENT AND SUBSTRUCTURES

structure as possible, preferably without movement joints and avoiding difficult shapes and complex profiles.

If joints cannot be avoided they will require very careful positioning, design and construction.

Diagram 5.2.10 shows where major movement joints may be needed.

See also the section on Frameworks.

Type A structures (CP102:1973)

Using a continuous imperious membrane

Most bitumen, plastic sheet or mastic asphalt tanking will accept some movement. It may split or crack by differential movement cracks in the supporting structure.

Unreinforced structures or a combination of brick/block and reinforced concrete will require movement control joints at the intersections with other materials (see diagram 5.2.11).

Type B structures

In type B structures (without membranes) cracks over 0.2 mm would lead to water penetration. The criteria for movement control are naturally more onerous. Cracks *under* 0.2 mm are considered acceptable, self sealing and water tight. As such cracks are unusual within 2.4 of a free movement joint, above this dimension concrete basement walls/floors, will require a system of reinforcement combined with movement joints to control cracking and keep this to 0.2 mm or below.

A high quality dense concrete is also required. 50% of drying shrinkage in this section of concrete (<300 mm) takes place in the first three months and can take years in thick sections. Up to a total of .04% (4×10^{-4}) (As an example a *5m slab* or wall $\times 4 \times 10^{-4}$ gives a crack width of 0.002 m, ie *2 mm*.)

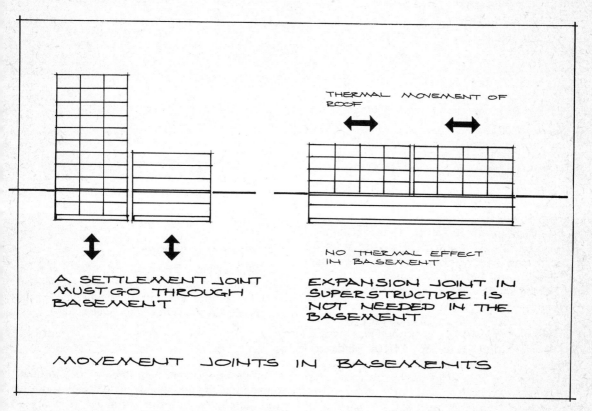

MOVEMENT JOINTS IN BASEMENTS

Diagram 5.2.10

DESIGN FOR MOVEMENT CONTROL IN BUILDING ELEMENTS

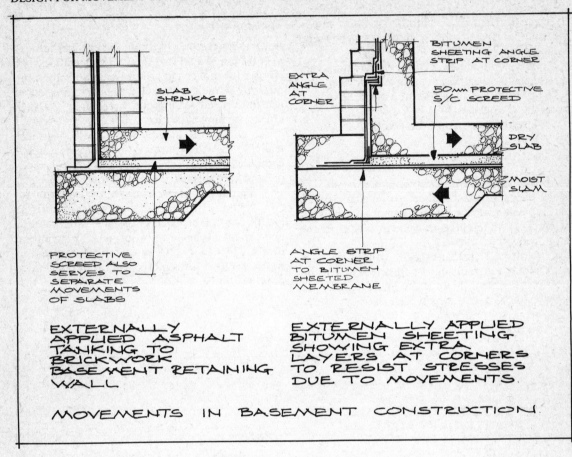

Diagram 5.2.11

Construction joints and movement joints

These are necessary every 5 m in walls and 10 m in slabs, with walls being cast vertically in one lift if possible.

See diagrams 5.2.12 and 5.2.13 for recommended details. BS 5337 lists five options for the layout of joints combined with various types of reinforcement to control and limit cracking.

CP 110 gives a percentage of reinforcement in basement walls to limit crack widths as follows:

Plain MS bars 0.30% horizontal plus vertical.
Deformed bars –0.25% horizontal plus vertical.

To limit initial cracking, CIRIA guide recommends for basement walls of 300 or less a provision of 0.5% secondary steel to gross cross sectional area of wall.

For base slabs a minimum of 0.17% steel in the top layer and 0.10 in bottom layer (and not less than 750 mm²/m in both cases). However bottom reinforcement is considered optional by others.

Reinforcement should follow all profiles and be well lapped and positioned.

Cracking due to reinforcement

As reinforcement provides restraint, if this is concentrated, it can lead to cracking. At vertical lapped junctions the reinforcement should be staggered to avoid a stress line developing.

Mix design

A compromise between workability for thorough compaction and a high aggregate/cement ratio to limit initial thermal and shrinkage cracking is necessary.

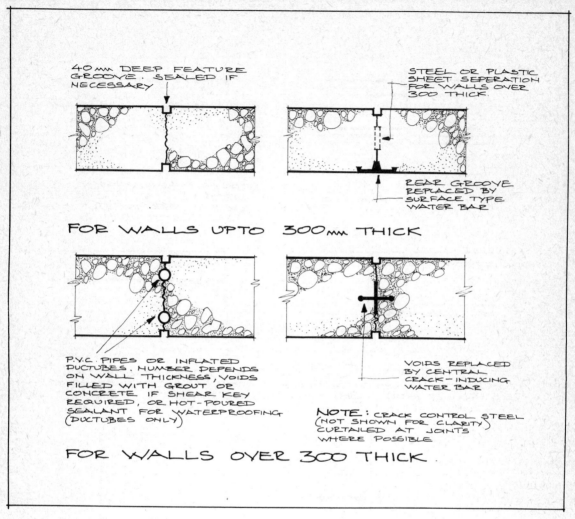

Diagram 5.2.12

5.3 EXTERNAL WORKS

EXTERNAL PAVED AREA: DESIGN CRITERIA

Pedestrian or vehicular loadings are the main variables.

Extrinsic causes of movement

Exposure conditions are similar in most cases, solar heat gain, frost action and sub soil water damage the main sources of extrinsic movement, together with

Differential settlement due to faults in the sub-base.

Intrinsic causes of movement

Mainly temperature changes in setting of concrete paving and initial shrinkage. Being external, due to unequal exposure (top to bottom) *warping* is possible and joints will mainly have to be provided to cover this possibility, although early protection of the concrete is also important.

DESIGN FOR MOVEMENT CONTROL IN BUILDING ELEMENTS

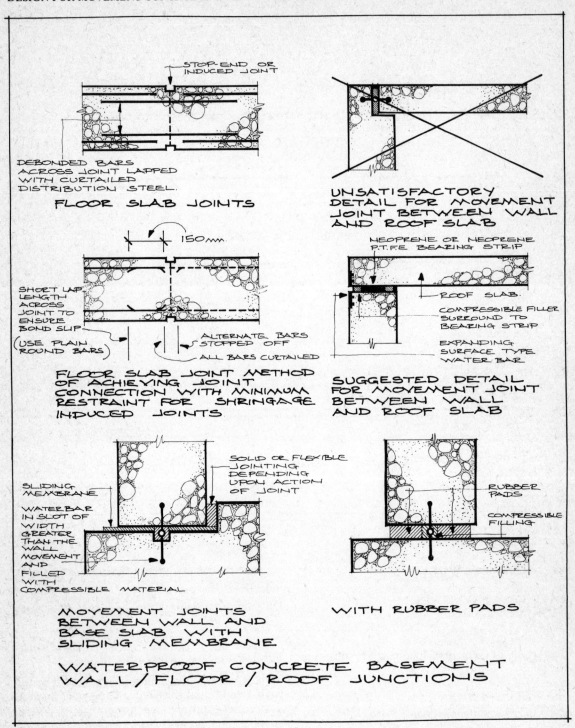

Diagram 5.2.13

Exposed concrete pavements and exterior exposed slabs

Diagram 5.3.1 shows alternative methods and layout of joints and reinforcement for external concrete surface slabs for pavements, courtyard, etc. The long strip method for large slabs can also be employed here.

Concrete slab and jointed surfacing materials

Where concrete is used as a base for other materials, early movement will still require a measure of control.

EXTERNAL FREE-STANDING WALLS

External walls have been covered in the section on walls, where the special differential exposure conditions of free standing walls are described.

STRUCTURAL FRAMEWORKS

SCOPE

The design of structural frameworks is usually the responsibility of the structural engineer, who will have to ensure that the framework can either resist or accommodate all movements which will develop, within acceptable limits. These limits have only recently been defined by the codes.

Relationship between frame and fabric

However, the building fabric cannot be considered separately from the frame and all movement control methods adopted for the frame will naturally require accommodation in the fabric, whether in panel walks, sheet or panel cladding and all forms of infill. It is essential that the design of the whole fabric is carried out in such a manner that all movements both in the frame and fabric are accommodated and particularly when movements are transmitted from one to the other.

Integrated design: structure and fabric

To achieve this, close collaboration between all concerned in the design of buildings is essential. It is hoped that the brief summary of theory and techniques in this volume will promote a better understanding of the principles being applied in the design of building structures by the various disciplines in the area of movement control.

The incorporation of structural features to provide stiffness design is well known. A better understanding of the principles involved will give fabric designers a better chance to incorporate these correctly at an early stage in design.

Finally, building fabric designers should rid themselves of their anachronistic faith in the rigidity of structures and gain awareness of the need to integrate the detail design of claddings and infill panels with the movements likely in the structural framework.

DESIGN CRITERIA AND FAILURE LIMITS

Structural stability/integrity

The stability of a structural frame can depend not only on the loads applied to individual structural members and resulting stresses, but on their deformation. These loads can cause

1 Unacceptable deflections in individual members.
2 Deformations in jointing mechanisms.
3 Displacement of mode points and *new* loading effects due to eccentricity and new moments.

Any of these can cause structural failure.

Diagram 5.4.1 shows the difference between local changes of shape (1) and in (11) displacement of mode points in a framed structure. Failure in (11) can occur if the joints as well as the members cannot distribute the loads *and* members are not capable of taking the *extra* loads due to displacement. Since these displacements are caused by shrinkage and creep phenomena as well as loading, these movements are now taken into account by the structural engineer.

Overall displacement of frameworks

Although stability usually depends on the strength of *individual* members, overall movement of the structure needs to be kept within limits as well.

Only recently have the codes indicated safe limits or limit states for the *overall* movement of structural frameworks. This is usually expressed as

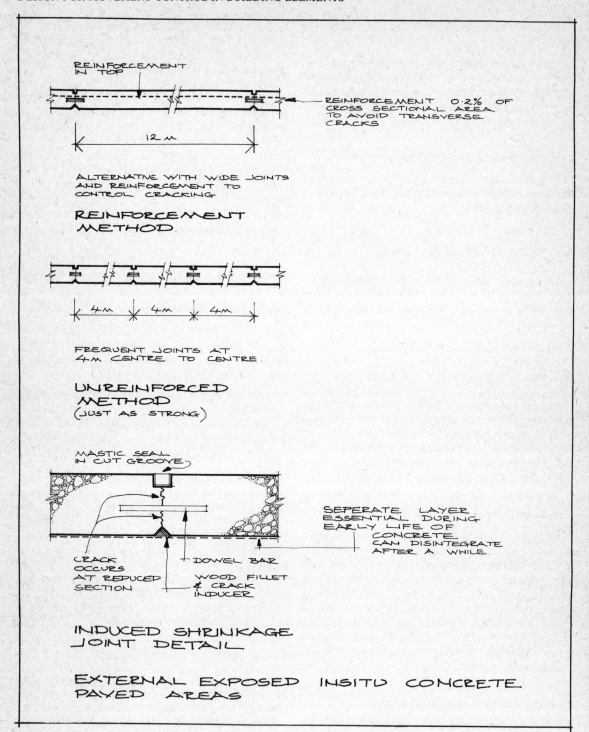

Diagram 5.3.1

STRUCTURAL FRAMEWORK

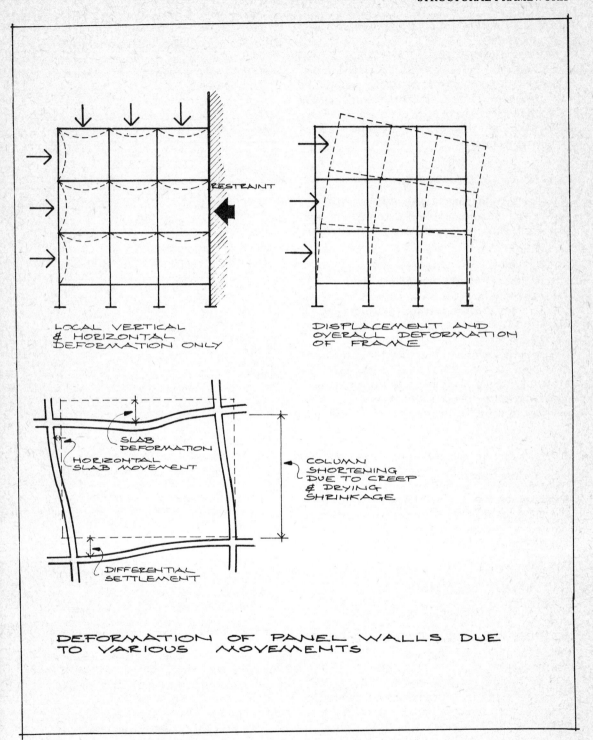

Diagram 5.4.1

DESIGN FOR MOVEMENT CONTROL IN BUILDING ELEMENTS

an angle or gradient from the vertical (see section on foundations).

Building Regulations do not state overall deflection limits but refer to Codes, which are now beginning to take this into account. It should be borne in mind that in order to provide a *rigid* structure a great deal of extra expense is necessary above purely avoidance of excessive stresses required for stability.

Structural design considerations

It is clear from the foregoing that structural design must take into account the deformations and details of joints and their displacement. There are two alternatives at joints:

1. deformation of two members can develop freely and without restraint ('pin' joint).
2. the restraint forces due to *differences in deformation* are safely absorbed and transmitted by the structural connection ('rigid' joint see diagrams 5.4.2 and 5.4.4).

The elastic deformation is then produced in all interconnected members and the connecting joint. Structural design takes these deformations into account in normal conditions of service.

Concrete/steel frames

In steel structures members are usually slender, and stress developed due to deformations can be more easily accommodated.

However in concrete structures where members are relatively larger in section and stiffer, more induced stresses due to deformations can develop. These may lead to cracking in the concrete.

DEFORMATION LIMITS

Deflection limits and cracking limits

These are normally based on the limits of the related fabric rather than the stability of the structure itself.

Cracking limits are also imposed for *durability in exposed* conditions. Beam/slab deflections are limited by the amount of movement which can be accommodated by various types of panel wall or finishes, ie the cracking strain limitation of infill panels, as shown in table 5.4.1.

Naturally, both deflection limits and cracking limits vary for different structural materials and the more stringent requirement would take preference, eg concrete casings to a more flexible steel framework.

Table 5.4.1 Deformation limits in structural frameworks:

Vertical deflections		
Beams	Steel beam	Span/200[1]
	Reinforced concrete beam	Span/250 or 30 mm[2]
	When supporting	
	(a) Brick/Block partition	Span/500 or 15 mm
	(b) Lightweight partition	Span/350 or 20 mm
Cantilever	When supporting cladding	Span 1/250–1/500 depending on cladding
	Visible limit	1/180
Lateral deflections		
Column	Sideways movement of multi-story building	Height/1000
	Frame diagonal bracing	1 in 600
	For racking failure panel wall in file	Height/500
	Single storey or low-rise frame.	Height/300
Mullion	As support to glazing (varies with glazing detail)	Span/175
Differential settlements Vertical	Floors and roofs	Span/250 to Span/500 depending on cladding and infill

1 Draft BSCP B/20 *The use of structural steel.*
2 BSCP 110 *Unified code for structural concrete.*

It is clear from the foregoing table that the allowable deflections and deformations will vary in accordance with the type of cladding used.

Some CLASP 1 and CLASP Mark 11 schools are recorded to have settled 1 part in a length of 270 and 1 in 170 without structural or cladding damage due to specially designed jointing and fixings of claddings (see diagram 5.4.3).

STRUCTURAL FRAMEWORK

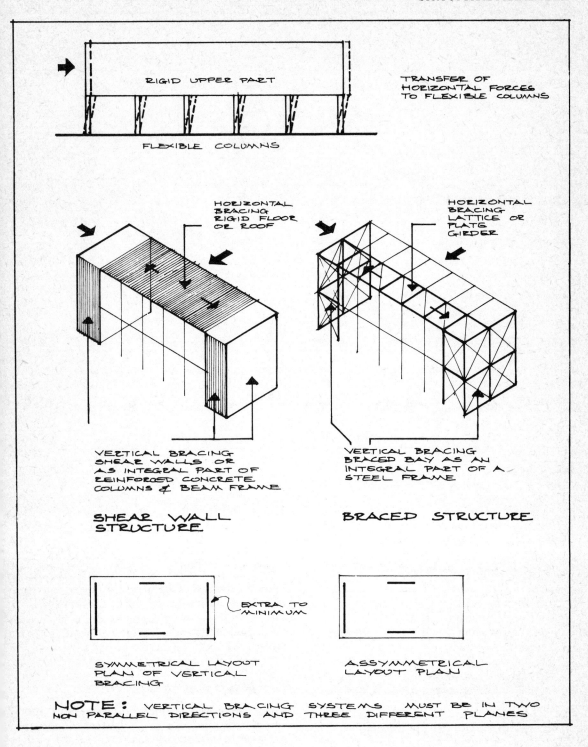

Diagram 5.4.2

DESIGN FOR MOVEMENT CONTROL IN BUILDING ELEMENTS

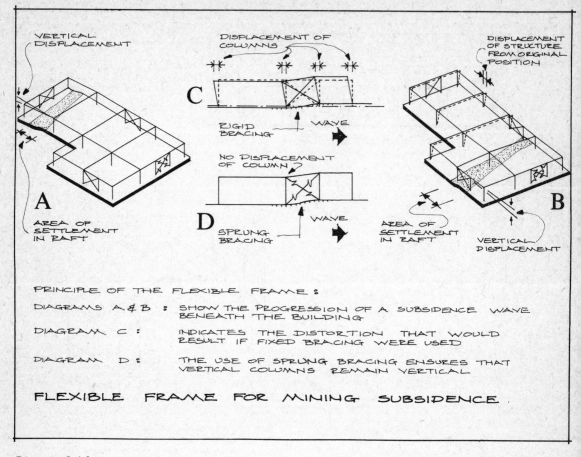

Diagram 5.4.3

Cracking limits in reinforced concrete (in frames or casings) based on durability criteria

Exposure grading (to BS 5337)	Crack width limit
A Exposure to moist or corrosive atmosphere and alternate wetting/drying	0.1 mm
B Continuous contact with liquid (ground)	0.2 mm
C General RC work in external but severe exposure (CP 110)	0.3 mm 0.004 × cover

(These widths are also visually acceptable)

For prestressed concrete limits are more severe: where cracking is permitted in certain members:

A normal conditions	0.2 mm
B severe conditions	0.1 mm

Since the stiffness of the structure depends on sizes of members joint design and reinforcement it is essential for the frame designer to be given the movement limitations, if other than standard, (see tables 5.4.1 and 5.4.2) at structural design stage.

Basis for deformation limits

The foregoing deformation limits have been derived mostly from the requirements of appearance, durability and effect on the rest of the fabric.

STRUCTURAL FRAMEWORK

Induced stresses by deformation and cracking

However, as referred to in earlier sections, deformation *imposed* on a member causes displacement which causes induced bending moments, whose magnitude depends on the stiffness and boundary conditions (ie degree of restraint: this is provided by the relative stiffness of adjoining members if rigidly jointed). In concrete structures, cracking can relieve induced stresses and if cracking limits are low the Engineer will have to take into account additional unrelieved stresses due to deformation which can be as high as those due to dead and live loads.

STRUCTURAL FRAMEWORK: CAUSES OF MOVEMENT AND EFFECTS

Table 5.4.2

Cause	Direction + type	Effect
Differential Settlement of foundations	Vertical and horizontal rotational movements	Cracking of columns undue deflection in floors/beams Cracking of floors Cracking of beams Disruption of cladding + partitions.
Loading (live, dead, wind)	Instant (elastic) Deflections and long term (creep)	Deflections, failure and cracking in columns Disruption of cladding
External temperature fluctuations (vary with degree of exposure and insulation and thermal capacity)	Horizontal Vertical	Roof slabs/decking Deformation of columns. Buckling in thin members. Distress in cladding/partitions mostly in exposed mullions
Internal temperature changes heating	Both horizontal and vertical	As above but mainly cracking in structural walls.
Moisture changes	Initial drying shrinkage horizontal shortening of long members. Alternate wetting drying.	Cracking at bearings, ribs and restraints. As above.

Causes of movement are also common to a particular structural material used for the framework as follows:

1. Steel framed structures
    (a) elastic deformations under load. (No time related deformation except dynamic due to vibrations or wind)
    (b) thermal movement
2. Concrete framed structures
    (a) drying shrinkage and moisture movements of concrete
    (b) elastic deformation under load (immediate)
    (c) creep of concrete under sustained load
    (d) thermal movement
    (e) differential movements and stress concentrations due to changes of stiffness.

MOVEMENT CONTROL METHODS
Movement reduction/restraint

Structural design

Design for stiffness by rigid members. This is the most expensive method (span depth/ratios).

Rigid joints or correctives or a combination of rigid and flexible joints.

Prevention of overall displacement by bracing/stiffening members or elements (see diagram 5.4.2 and 5.4.4).

Protection

Protection of frame by insulation, particularly where exposed to solar gain.

Choice of material and workmanship

Mix design and workmanship in concrete to reduce early age movements.

Selection of timber species with low movement characteristic and moisture content control.

Restraint by reinforcement in concrete structures as given in A 3.1 of CP 110.

In reinforced concrete cracking due to strains by loading depends on the following factors:

1. proximity to the point considered of reinforcing bars perpendicular to the cracks.

DESIGN FOR MOVEMENT CONTROL IN BUILDING ELEMENTS

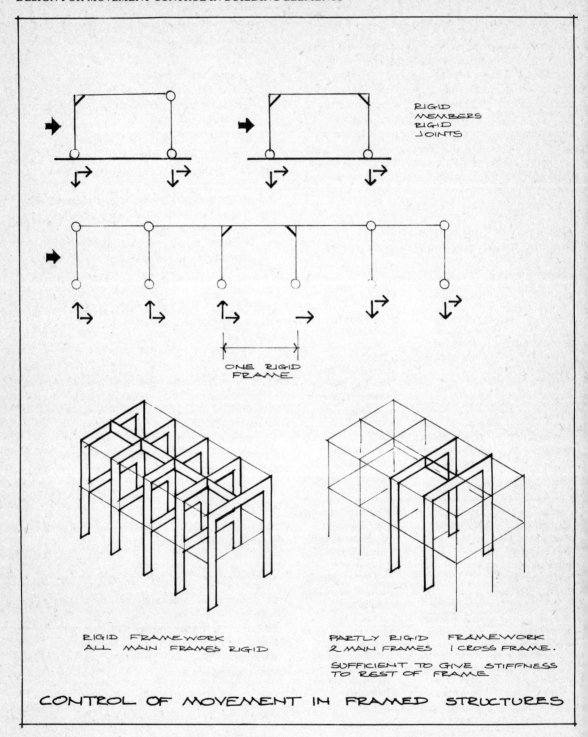

Diagram 5.4.4

2 Proximity of neutral axis to point considered.
3 Average surface strain at the point considered.

CP 110 in paragraph A 3.2 gives a crack width formula giving the relationship between these 3 variables and the crack widths likely to occur. Provided the final strain of the member, *NB corrected for additional shrinkage and creep strains* is known, then the crack widths can be controlled by a suitable area of reinforcement, suitably positioned, and within the crack width limits given in table 5.4.2.

However, there is a limit of the amount of strain which can be accommodated by restraint, ie reinforcement, within acceptable crack width limits or deformation/deflections. Unsightly cracking, bowing and differential movements will then need to be controlled by accommodation of the movements as follows:

Time related movement control

The initial movements in the frame, ie deflections and early age movements in concrete can be allowed to take place *before* sensitive facings and claddings are applied. (CP 297 1972 *Precast cladding*, clause 3.7.1 and 3.7.3.1 recommends fixing of cladding after structure is complete.)

ACCOMMODATION OF MOVEMENT

Where it is not possible to restrict or prevent excessive movements, these have to be allowed to take place without damage to the structure or fabric, ie they must be accommodated by:

1 Reduction of restraints on the free movement of the structure.
2 Subdivision of structure into suitable separate lengths or portions separated by the appropriate movement joints.

Judicious placing and construction of these joints is essential if the movements are to be controlled, ie if all portions are able to move without causing induced stresses in members of sufficient magnitude to cause cracking or unacceptable deformations.

Movement joints in framed structures

The structural framework itself is often not exposed to the extreme external temperature fluctuations and will itself not suffer from undue movements.

However, frameworks can be affected by differential movements in other elements, eg foundation settlement or expansion/contraction of the roof. See diagrams 5.4.1, 5.4.3, 5.4.5 and 5.4.6.

In steel and timber frames, where the roof or floors are often not integral or continuous with the structure this is not such a serious problem as in continuous reinforced concrete structures where all stresses are transferred. Also where parts of the structural frame may be exposed externally, eg columns.

However the need to accommodate large movements in roof slabs or the risk of differential settlements usually requires a movement joint in the frame itself with portions independently supported, as sliding bearings are very difficult to construct and are prone to failure.

Steel frame structures

Major movement joints in the frame are usually only needed due to the possible movements of other elements as described earlier in which case a double column and beam structure is usually necessary, since sliding bearings are difficult to construct. However, some tolerance on movement of supported slabs are possible, (see diagram 5.4.5).

Flexible steel frames, with pin joints which can accommodate movement and flexible bracing are feasible, provided the cladding can be suitably designed to accommodate this movement.

Timber framed structures

The same principle of the flexible frame would apply to timber frame structures again with the proviso that the cladding and roofing systems must also be flexible to allow for the anticipated movements.

In concrete framed structures

Unlike steel or timber structures, concrete continuous *in situ* structures result in transfer of movement induced stresses unless joints are formed in the structural framework.

DESIGN FOR MOVEMENT CONTROL IN BUILDING ELEMENTS

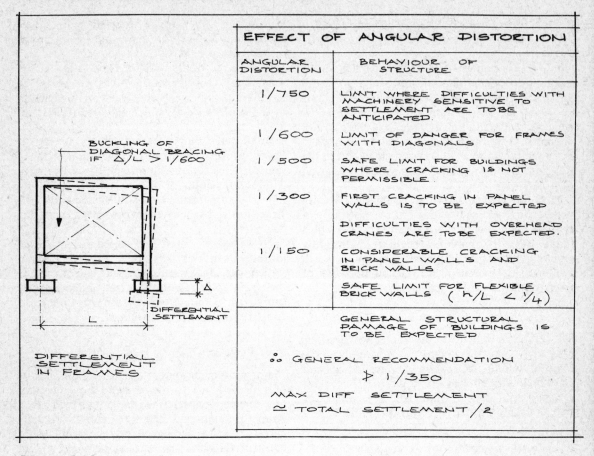

Diagram 5.4.5

MOVEMENT JOINTS: CONCRETE STRUCTURES (see diagram 5.4.6—5.4.9)

For concrete structures, CP 110 part 1, 1972. The structural use of concrete and CP 116:1969 the structural use of precast concrete both define the following types of joints based *on their function*.

Types

1 *Contraction joints*: No initial gap but deliberate discontinuity to permit contraction.
Complete both concrete and reinforcement are interrupted.
Partial where concrete only is interrupted.
2 *Expansion joints* (concrete and reinforcement discontinuous) to permit movement either expansion or contraction.
3 *Sliding joint* (concrete and reinforcement discontinuous) with special provision to permit horizontal movement.
 As these involve transfer of loading, ie bearings they are difficult to form and liable to failure of ribs and supports.
4 *Hinged joint* to permit relative rotation of members.
5 *Settlement joints*

 Lasly joints may be required to fulfil more than one function.

Details

The detailing of joints in the structural frame has to take account of load bearing requirements and becomes the responsibility of the structural

STRUCTURAL FRAMEWORK

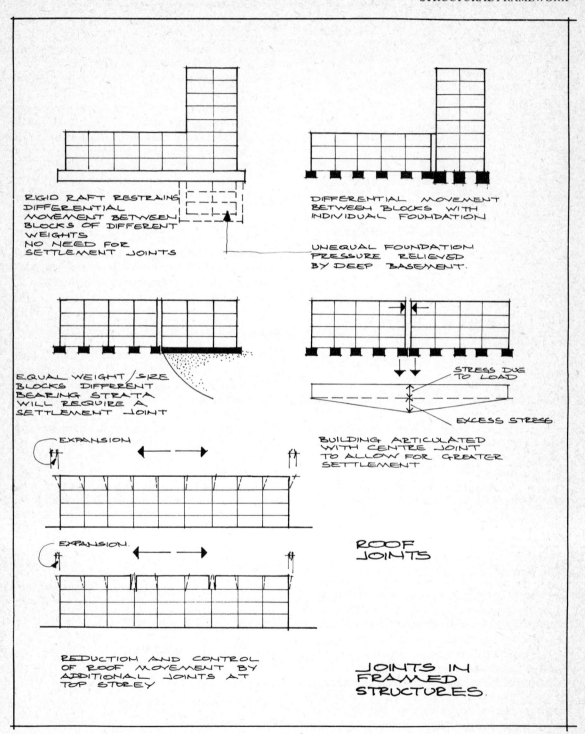

Diagram 5.4.6

engineer (see diagrams 5.4.7 and 5.4.8). However, waterproofing, fire stopping and the requirements of finishes should involve all those responsible for fabric design and detailing. In precast concrete or precast and *in situ* combined frameworks such joints, together with the construction joints or assembly joints in precast concrete can be important and visible features of the fabric (see diagram 5.4.9).

Restraint: The effect of adjacent elements

Where surrounding or adjoining elements provide restraint, then the free linear movement of the wall is restricted and deformations other than straight linear movement, ie bowing buckling can be caused.

5.5 WALLS: BRICK AND BLOCK MASONRY

SCOPE

This section covers walling in brickwork and blockwork.

Both clay, calcium silicate and dense or lightweight concrete are considered. These materials vary considerably in their movement characteristic.

Both loadbearing walls and non loadbearing panel or partition walls require examination.

MOVEMENT CONTROL CRITERIA

Basic failure limits

The basic requirements of stability, weather exclusion, integrity and durability can be seriously impaired by deformations and stresses caused by dimensional changes. As with foundations, the degree to which these movements are controlled in walls is determined by the need to preserve the functional performance of the element or elements of which the wall forms part. Failure limits in the walling may be determined by failure limits in more susceptible subsidiary elements, ie cladding, plaster or rendered finishes. (See table in diagram 5.1.2 and diagram 5.5.1.)

Subsidiary components

The acceptable movement limits of various facings and claddings particularly wet bedded and rigid jointed units are covered in the section on finishes.

The allowable limits in the back-up walling, for finishes such as plaster and renderings are very small possibly not greater than 0.02%, well below the likely shrinkage movement in blockwork, or the differential movement between walling and other components, eg beams and columns.

Facings and cladding

These claddings or finishes, if subjected to movement in excess of their stress resistance (of the facing or its fixing) may either crack, become detached due to failure of fixings or bedding.

The overall performance of the wall may then be seriously affected.

Failure limits: appearance

Since good appearance is an essential part of wall performance when the walling is used as a facing element it is possible that any *visible* cracking may be unacceptable, even though there is no serious loss of stability.

The degree of control required will depend on these acceptable failure limits.

Table 5.5.1 gives the limits of normally visible crack widths.

Where brickwork or blockwork are used as a facing material in walling table 5.5.1 (taken from CIRIA, technical note 107 *Design for movement in buildings*) is a useful guide.

Table 5.5.1

Degree of damage	Description of typical damage	Approximate crack widths
Very slight	Cracks visible on close inspection only	<1 mm
Slight	Cracks easily filled	<5 mm
Moderate	Doors and windows stick weather tightness may be impaired	5–15 mm
Severe	Extreme repair work required visible distortion bulging of walls	>15 mm

WALLS: BRICK AND BLOCK MASONRY

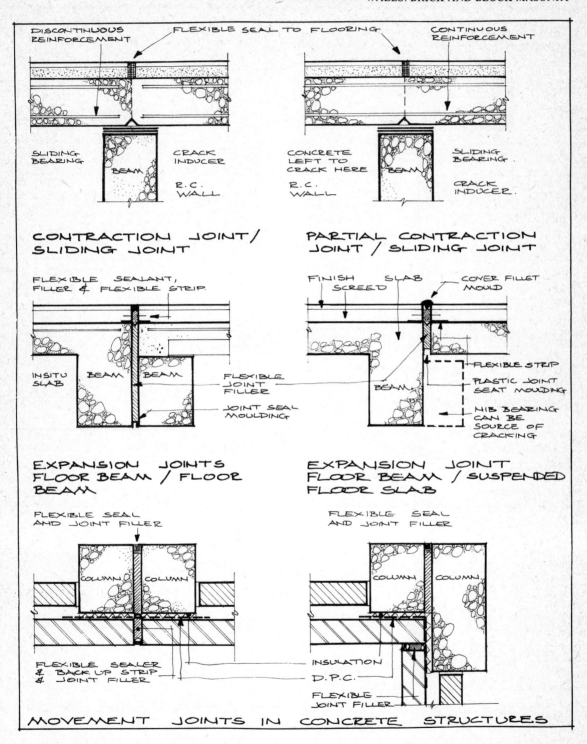

Diagram 5.4.7

DESIGN FOR MOVEMENT CONTROL IN BUILDING ELEMENTS

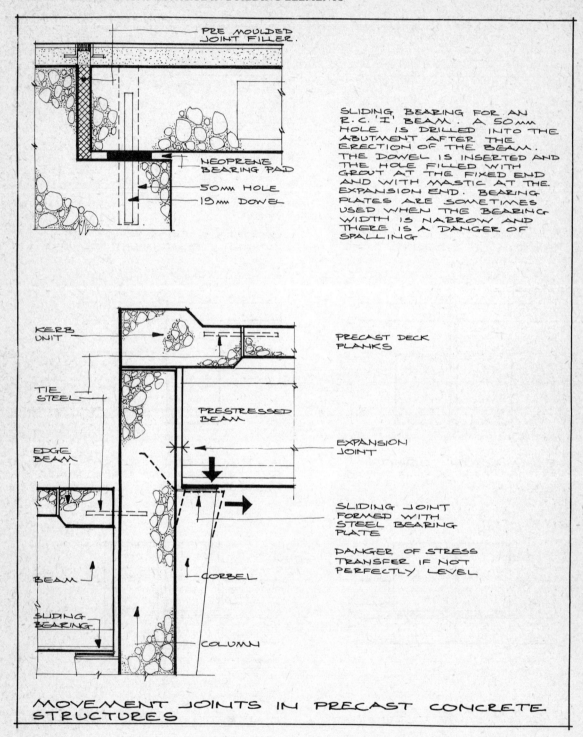

Diagram 5.4.8

WALLS: BRICK AND BLOCK MASONRY

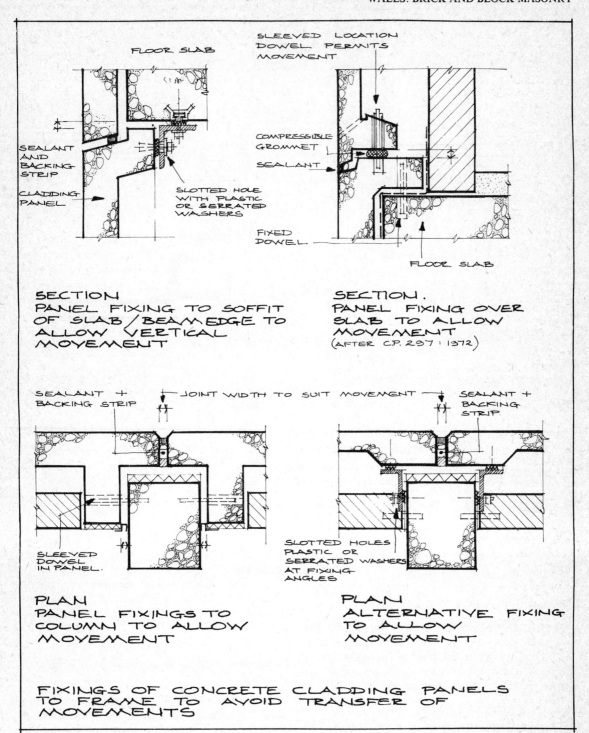

Diagram 5.4.9

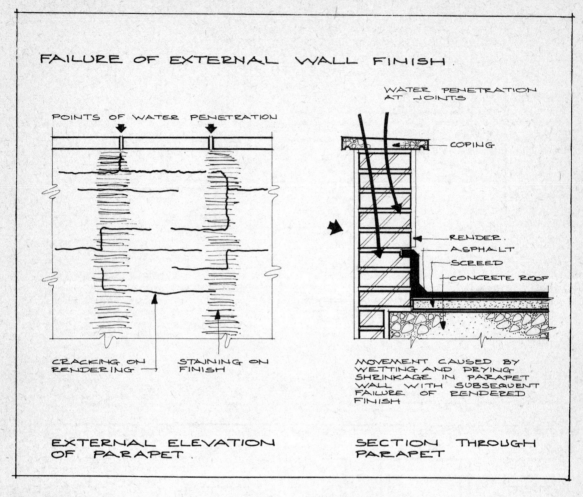

Diagram 5.5.1

Settlement of supporting foundations or structures can cause sagging, hogging in the walling.

Settlement of supports

The section on foundation gave the criteria for allowable deformations in the walling both as a differential between one part of the walling and another by means of the maximum angle of deflection and of the overall movement of the wall for stability.

ASSESSMENT OF MOVEMENTS IN WALLS

Before appropriate methods of control can be decided the causes, magnitude, locations and likely affects of movements have to be identified and assessed.

The type and location of movements depend on the general location of the walling, the adjoining elements as well as the constituent materials of the walling.

Disposition and location of walling traditional solid masonry

In order to assess the likely movements which may occur it is necessary to take into consideration the general disposition, shape, form and location of the walling as well as the surrounding construction.

WALLS: BRICK AND BLOCK MASONRY

Table 5.5.2 (taken from *Illustrated Introduction to Brickwork Design*, Nov. 1975 by T L Knight issued by BDA) shows typical total movements for a 10 m length of wall.

Table 5.5.2

Movement cause	Movement over 10 m lengths	Damage category
Thermal (50°C)	3 mm	Slight
Drying shrinking of calcium silicate brickwork	−1.35 m	Slight
Total clay moisture expansion after 15 days	+4.5 m +2.5–3 m	Slight/moderate Slight
Combined thermal and moisture expansion 7		Moderate

In older buildings the form of the masonry was usually massive, ie a low ratio of thickness to height and length, this reduced the stresses due to changes of dimension by greater distribution of stresses.

Smaller openings also reduced the severity of exposure as will be discussed later and the incidence of weak or critical points.

Masonry in thin multi-skin layers

Many buildings today use masonry in large areas of minimal thickness as relatively thin layers in cavity or diaphram walls (see diagram 5.5.2). This results in smaller cross sectional areas subject to larger temperature and moisture fluctuations, deformations and resulting stresses.

There is also a differential in the exposure conditions between the layers (see diagram 5.5.2). This is accentuated by the new technique of cavity insulation, which subjects the outer skin to higher solar temperatures and frost action.

The maximum temperatures reached in south facing masonry are given earlier in the general section on sources and estimation of movements.

LOCATION: RESTRAINT

Since unrestrained movement is not as serious in effect as the stresses caused by movement which is restrained, the form of the walling and its location in relation to surrounding elements, buildings and all sources of restraint and movements imposed by adjoining elements are critical.

Building shape

What is not so often realised is that a change of direction is in itself a powerful restraint, ie the corner of a building or offset in the facade (see diagram 5.5.3).

Change of direction/corners

Again, contemporary buildings tend to be complex in plan shape with many changes of direction. It is essential to recognise this problem and to place the appropriate movement control joints in the walling, as shown in diagram 5.5.4 in a number of examples.

Articulation of the building at the points where changes of shape and direction occur can avoid the need for wall joints and HIDE Joints. It is also useful to provide stability to the end sections of wells where joints might weaken the overall stability of the building (see diagram 5.5.5).

The effect of adjacent elements

Where surrounding or adjoining elements provide restraint then the free linear movement of the wall is restricted and deformations other than straight linear movement, ie bowing buckling can be caused.

Imposed stresses due to movements

More serious effects can result from actual deformations in the adjacent elements and surrounding construction. Movements and stresses are imposed on the walling, often when there is no inherent cause of unacceptable movement in the walling itself.

Serious local stresses far in excess of the limitations of the walling can result, with subsequent failures.

Diagrams 5.5.6–5.5.9 show the effects in more detail indicating the points where the movement stresses are imposed and in some cases the distribution of the effects.

DESIGN FOR MOVEMENT CONTROL IN BUILDING ELEMENTS

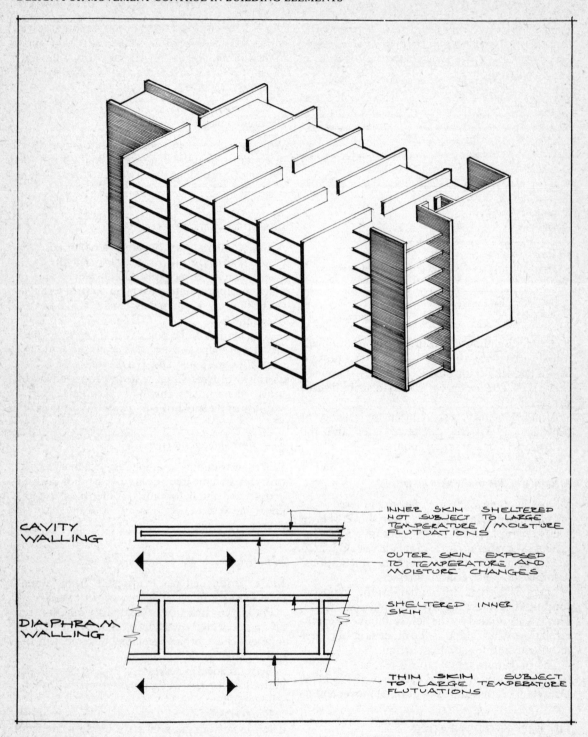

Diagram 5.5.2 Based on diagram page 3 SCP tn 4

WALLS: BRICK AND BLOCK MASONRY

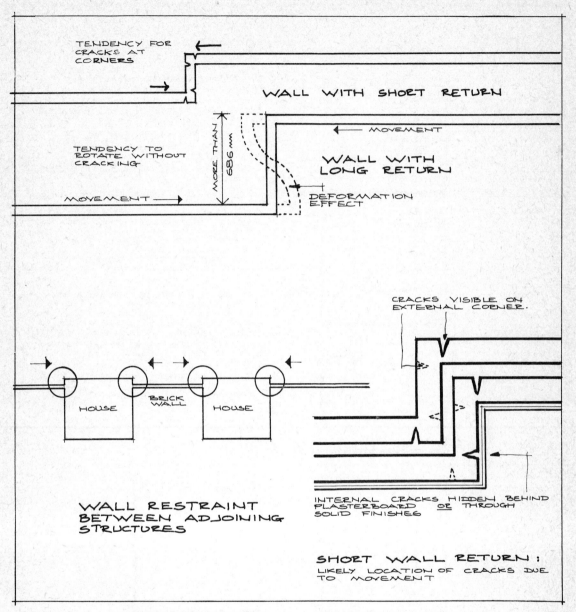

Diagram 5.5.3 Based on diagram page 23 SCP tn 4

DESIGN FOR MOVEMENT CONTROL IN BUILDING ELEMENTS

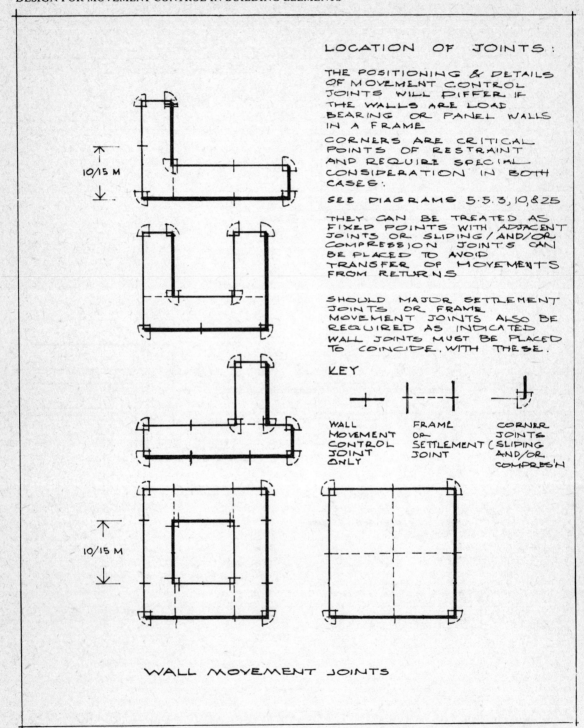

Diagram 5.5.4 Some typical joint layouts. Every building will require individual assessment. Based movement in service as shown in the following

WALLS: BRICK AND BLOCK MASONRY

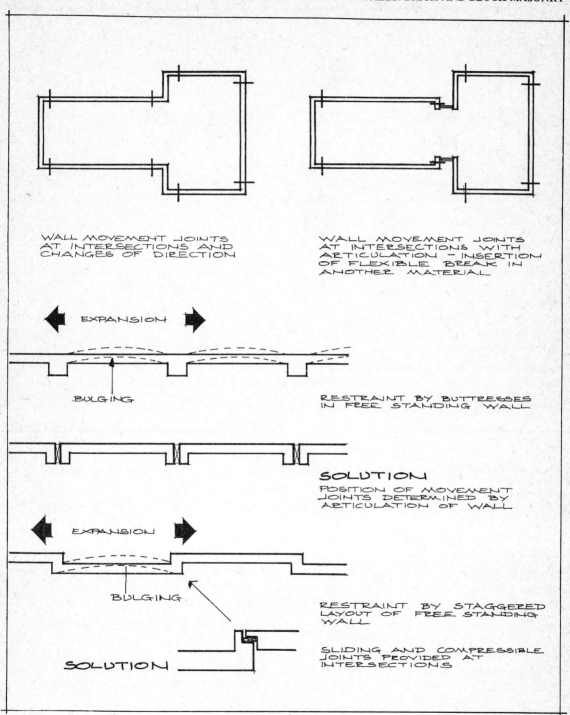

Diagram 5.5.5

DESIGN FOR MOVEMENT CONTROL IN BUILDING ELEMENTS

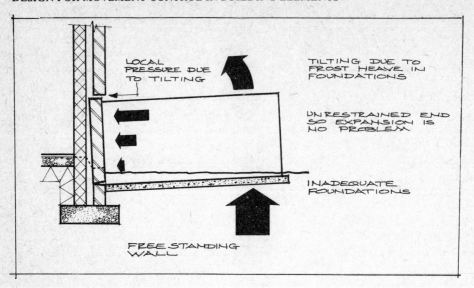

Diagram 5.5.6

Differential movements

Finally, adjoining elements may in themselves not be subject to excessive movements, but being constructed in dissimilar materials from the walling cause differential movements which can cause failure in the walling.

DEFORMATION IN ADJOINING ELEMENTS

1 Deformations due to loading

Movements and deflections of supports to the walling:

Beams, particularly cantilevers, floor slabs, shortening of columns.

Deflections to structure over non-loadbearing walling.

Local deformation of slabs and beams on bearings due to curvature.

Racking of frames.

2 Deformations due to other causes

Thermal movement:
Roof slabs: horizontal movement.
Frameworks: differential vertical movement.
Drying shrinkage (induced).
Curvature in beams/slabs/columns.
Creep in Structural frameworks.

SURFACE CHARACTERISTICS: SEVERITY OF EXPOSURE

The importance of aspect has already been stressed.

Aspect

The colour and reflectivity of the wall surface will affect the solar absorption and resulting service temperature range.

Surface characteristics

Naturally the designer can mitigate the effect of exposure to solar radiation by selection of suitable materials. The placing of thermal insulation immediately behind the exposed layers of the wall can greatly increase the heat build up and also the temperature drop due to lack of heat from the interior.

Table 2.2 from BRE Digest 228, part 2 shows the variation of service temperatures which can result from the surface characteristics of the wall.

WALLS: BRICK AND BLOCK MASONRY

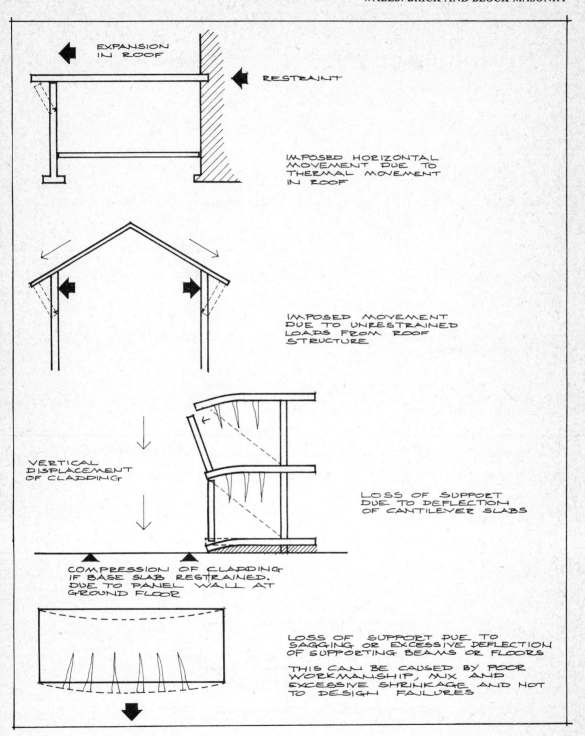

Diagram 5.5.7

DESIGN FOR MOVEMENT CONTROL IN BUILDING ELEMENTS

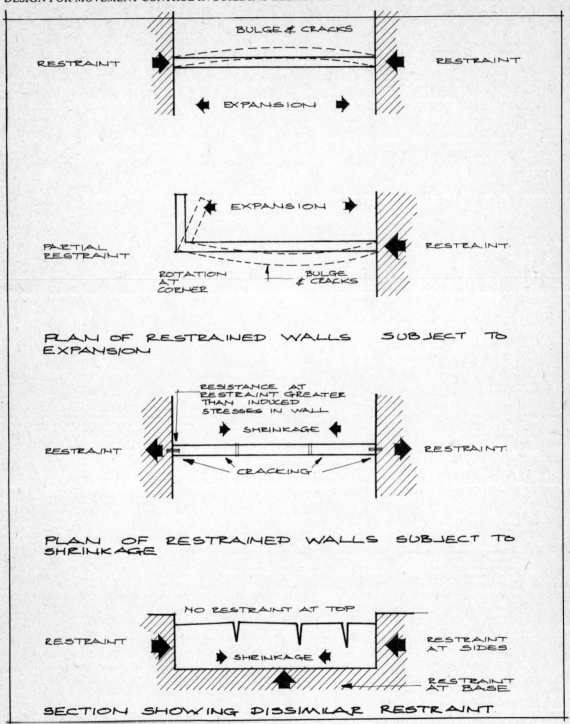

Diagram 5.5.8

WALLS: BRICK AND BLOCK MASONRY

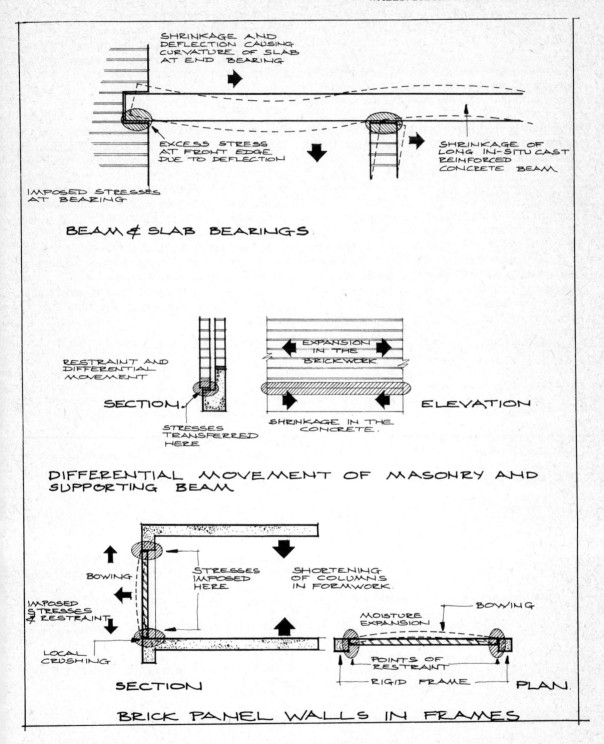

Diagram 5.5.9

DESIGN FOR MOVEMENT CONTROL IN BUILDING ELEMENTS

LOCATION OF CRITICAL POINTS IN WALLING

As failure is most likely to occur at these points the designer should identify these points as soon as possible so that measures can be taken to control the effects.

Failure is most likely to occur at points where stresses due to movement become concentrated (see diagram 5.5.10).

Area of stress concentration

Failure due to movement is caused when the stresses induced by or caused by the linear movement of a wall are restrained or restricted causing tension stresses which lead to cracking. If these stresses are concentrated in smaller areas the likelihood of failure in these locations is greater. The design must take this into account and extra strength by reinforcement or the provision of movement joints is essential at these points.

These points mainly occur in the following locations (see also diagram 5.5.10).

movement likely to be encountered have to be identified and their location, extent and magnitude estimated. The combined effect of several types of movement may have to be considered.

Prediction of effects

The effects of such movement can then be predicted. As described in the general section on the estimation of movements and their effects, the resulting deformations depend on many factors and in most cases due to inequalities of composition, exposure between faces, or restraint conditions, some deformation other than purely linear will occur, ie twisting, buckling, bowing. Identification of weak and critical points as described earlier enables the designer to determine the most likely location of failures.

The brick and block walling movements can be identified and classified under the types given in table 5.5.4.

1	Changes in wall dimension	Wall height
		Wall thickness
		Local reductions in thickness due to vertical or horizontal chases or bearings
2	Changes in wall direction	Corners and short returns
3	Changes in wall area	At the smaller areas between openings or adjacent to openings
4	Changes in wall stiffness or material	Adjacent to piers or other stiffeners or points of greater flexibility
5	Areas of greater exposure	Parapet walls
6	At points of greater restraint than general run	Junctions with crosswalls, butt ends
7	At points of weaker restraint than general run due to loss of bond	Over damp proof courses and continuous flashings
		(*Note* These can form a useful slip plane to control movements and save differential movement transfer)
8	At junctions with cladding or other elements/components	Changes of cladding, eg string course

IDENTIFICATION OF CAUSES OF MOVEMENT AND PREDICTION OF EFFECTS

Procedure: identification: estimation and control

Following the examination of the form disposition and surface character of the walling, the types of

Identification and location of differential movement in walls

Types of differential movement

Differential movement of the walling and applied facings and cladding are common and can cause

WALLS: BRICK AND BLOCK MASONRY

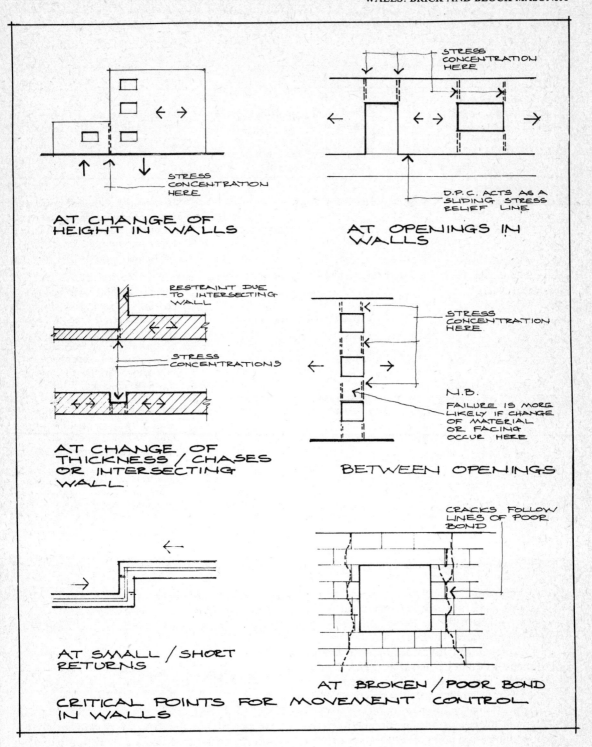

Diagram 5.5.10

DESIGN FOR MOVEMENT CONTROL IN BUILDING ELEMENTS

Summary of causes and types of movement in walls
(see diagram 5.5.11–5.5.14)
Table 5.5.4

Causes	Type of movement	Location and effect
Extrinsic movements		
Due to loading	Deflection of supporting slabs or beams	Bowing or sagging of wall. Vertical differential movement
	Creep in columns	
Non load related	Thermal movement shrinkage of supported slabs, roofs	
Intrinsic movements		
Due to loading	Elastic deflection	Shortening, buckling or bowing if load or slenderum excessive
Not due to loading	Thermal movement in brick or blocks	Linear expansion or contractions other deformation only if restrained or differential exposure
Not due to loading	Moisture movements: irreversible	Initial irreversible expansion of clay bricks: bowing if restrained.
	Alternate wetting/drying	Initial curing shrinkage of concrete bricks or blocks, drying of concrete *blockwork due to excessive initial absorption/ saturation. Shrinkage of mortar shrinkage cracking.
Other	Physical and chemical changes (a) Corrosion	Expansion of fixings to masonry causing local imposed stresses. Loss of fixing due to loss of restraint, eg wall ties or support, eg lintels.
	(b) Sulphate attack	On mortar: expansion of wall joints and walling, usually greater vertically
	(c) Carboration	Shrinkage of porous portland cement products, eg lightweight concrete blockwork.

* This is the main cause of movement in walls constructed of lightweight concrete blockwork.

serious failure of the cladding and facings especially as these are usually very sensitive to movement, either in the material itself or in the fixing mechanism (see cladding section).

A 15°C temperature difference between facings and the back up walling behind has been shown to be responsible for the failure of thin stone facings.

This type of failure is particularly serious when the facing is *applied over* a point of locally concentrated stress, eg at the junction between the back up walling and a point of restraint, eg a frame (see diagram 5.5.16).

Stresses due to differential movement in walls
Differential movement in multi-layer walls

(see diagram 5.5.17)

ESTIMATION OF EFFECTS

Estimation of magnitude

After identifying the type of movement to be expected and its direction and location is determined, its magnitude should then be predicted.

Table 5.5.5 Causes of movement related induced stresses

Causes	Type of movement	Location and effect
Differentials immaterial composition/location/ exposure	Differential movements	Between areas of walling subject to different exposure (see diagram 5.5.15). Between masonry and adjoining concrete or other structures in dissimilar materials
		Between bricks of different types
		Between brickwork and damp proof courses or other built in sources of low resistance to horizontal movements
		Between brickwork and blockwork skins in multi layer walls
		At junctions with existing or other rigid structures

Material properties

This usually requires an assessment of likely conditions, eg moisture, temperature changes, followed by a calculation based on the movement properties of the constituent wall materials. The reader is referred to the data given in the general section for each material and to section 2, movement estimation.

Movement prediction

Prediction of the likely movement is bound to be only approximate. Professor De Courcey of Dublin University, a leading authority in the subject, claims only to predict 'trends'. This is because actual exposure is difficult to forecast, accurate data is hard to obtain, the interaction of various effects is hard to predict.

The building will always find its own weak points even if the designer has not done so as has been stressed before.

Finally as movements are time related, the time of construction which cannot always be predicted also imposes a significant variable factor. However it is still important to form as true a picture as possible and then exercise judgement in the adoption of preventive or control methods.

SELECTION OF MOVEMENT CONTROL METHODS

Alternative control strategies

There are three broad strategies, as in the design for other elements to control the effect of movements in walling.

1 *Accommodation of movements*
by separating the wall into sections so that each is free to move without detriment to the wall performance or of adjoining elements.

2 *Mitigation or reduction of movement or its effects*
by protection from exposure to moisture or temperature fluctuations (eg reflective surfaces)

By selection of materials less likely to move.

By workmanship: protection of blockwork from excessive moisture absorption.

By adoption of design details which separate materials with dissimilar properties.

3 *Prevention of movement*
By reinforcement (this can usually only apply locally or for small areas of finishes).

MOVEMENT CONTROL METHODS IN WALLS

Alternative 1: accommodation
Alternative methods of accommodating the movements rather than reducing or preventing the movements are

1 Creation of discrete panels or separate sections by design features insertion of components to act as a break.

DESIGN FOR MOVEMENT CONTROL IN BUILDING ELEMENTS

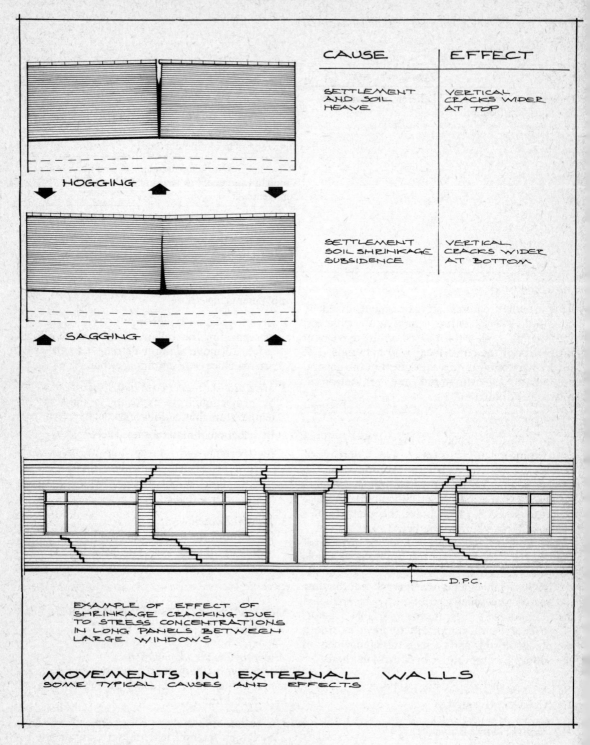

Diagram 5.5.11 Based on diagram page 11 SCP tn 4

WALLS: BRICK AND BLOCK MASONRY

	CAUSE	EFFECT
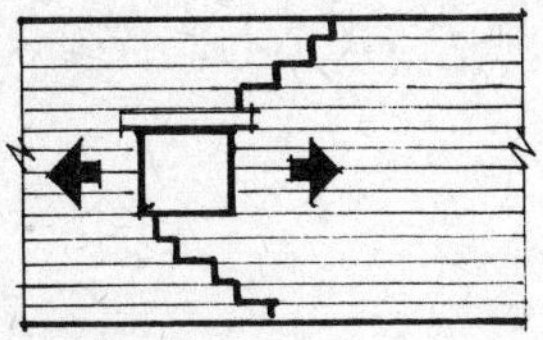 **SHRINKAGE OF CONCRETE BLOCKWORK**	Moisture shrinkage of masonry failure at weak points.	Diagonal and vertical cracks
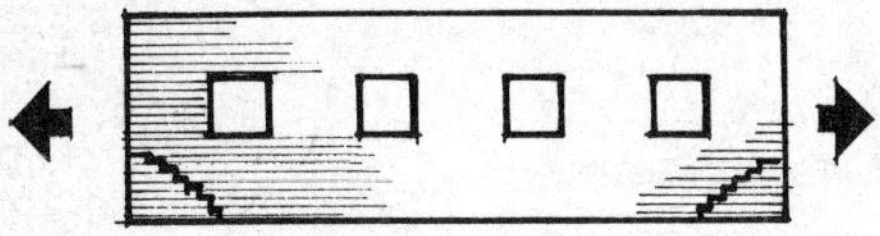 **EXPANSION OF CLAY BRICKWORK**	Thermal and initial expansion of brickwork	Diagonal cracks at points of restraint (corners)
 **SULPHATE ATTACK**	Excessive damp in brickwork leading to sulphate attack on joints	Horizontal cracks at irregular centres
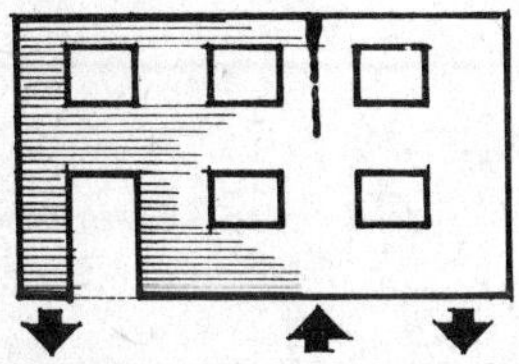 **FOUNDATION MOVEMENT**	Sulphate attack on foundations. Frost heave. Swelling of clay at centre.	Vertical cracks wider at top
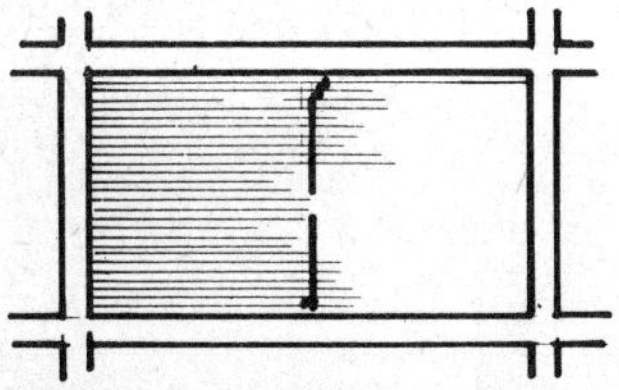 **RESTRAINT TO BRICKWORK EXPANSION**	Moisture expansion of brick panels	Bowing and even vertical cracks.

MOVEMENT IN EXTERNAL WALLS.
SOME TYPICAL CAUSES AND EFFECTS

Diagram 5.5.12

DESIGN FOR MOVEMENT CONTROL IN BUILDING ELEMENTS

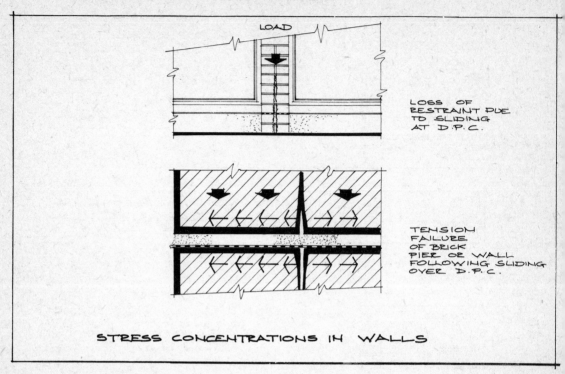
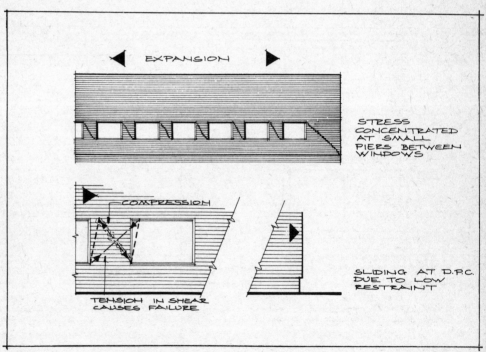

Diagram 5.5.13 Based on diagram page 21 SCP tn 4

WALLS: BRICK AND BLOCK MASONRY

	CAUSE	EFFECT
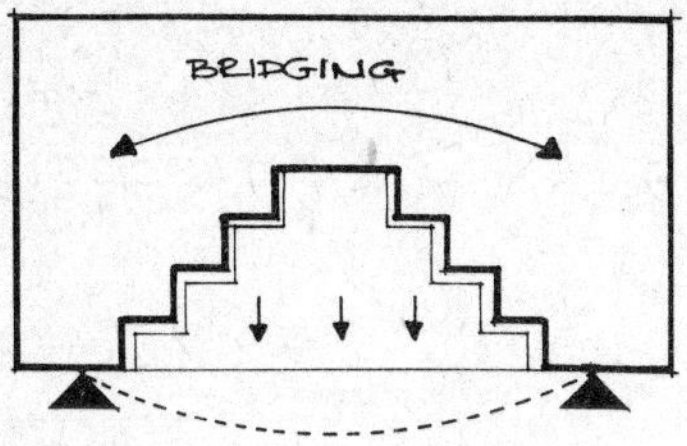 *BRIDGING* (diagram)	DEFLECTION OF FLOOR SLAB SUPPORT	CRACKS MAINLY AT BASE HORIZONTAL AND STEPPED DUE TO BRIDGING
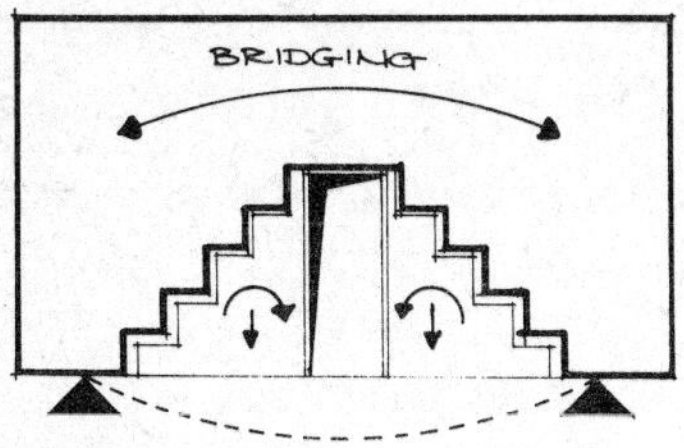 *BRIDGING* (diagram with opening)	DEFLECTION OF FLOOR SLAB SUPPORT	AS IN ① BUT BRIDGING OVER OPENING
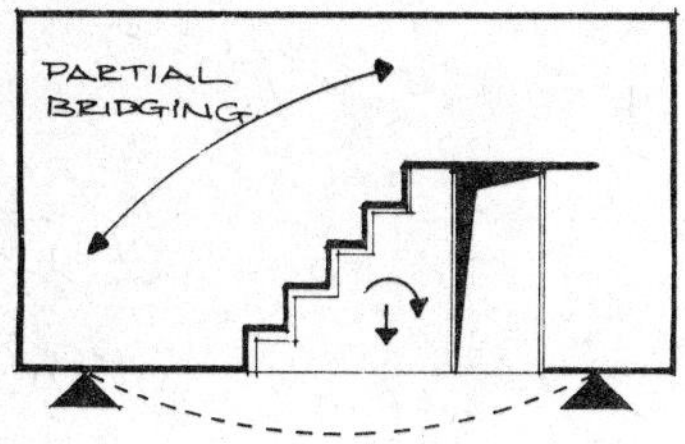 *PARTIAL BRIDGING* (diagram)	DEFLECTION OF FLOOR SLAB SUPPORT	AS IN ① & ② BUT ASSYMMETRICAL BRIDGING
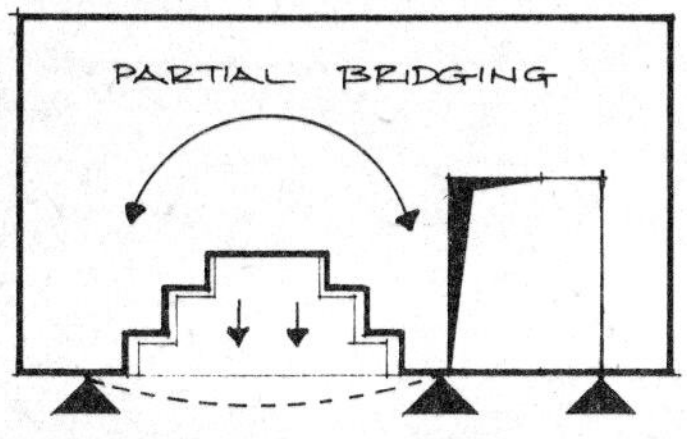 *PARTIAL BRIDGING* (diagram)	DEFLECTION OF FLOOR SLAB SUPPORT	AS ABOVE BUT ASSYMMETRICAL BRIDGING
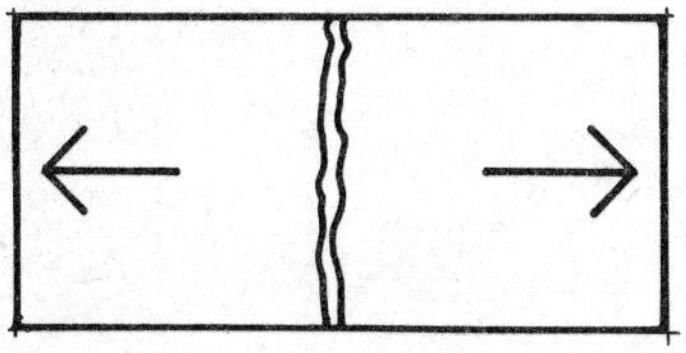 *Vertical crack with arrows*	DRYING SHRINKAGE	VERTICAL CRACKS IF ENDS RESTRAINED
	CAUSE	EFFECT

MOVEMENT IN INTERNAL WALLS.
SOME TYPICAL CAUSES AND EFFECTS.

Diagram 5.5.14

DESIGN FOR MOVEMENT CONTROL IN BUILDING ELEMENTS

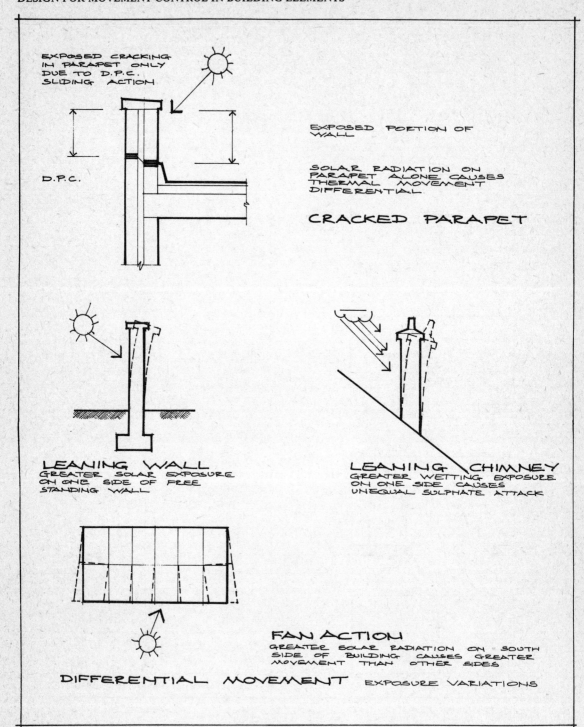

Diagram 5.5.15

WALLS: BRICK AND BLOCK MASONRY

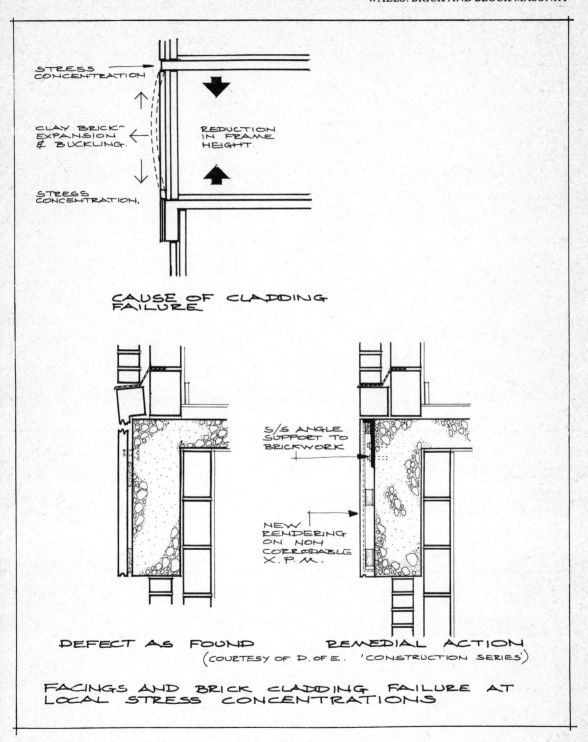

Diagram 5.5.16

DESIGN FOR MOVEMENT CONTROL IN BUILDING ELEMENTS

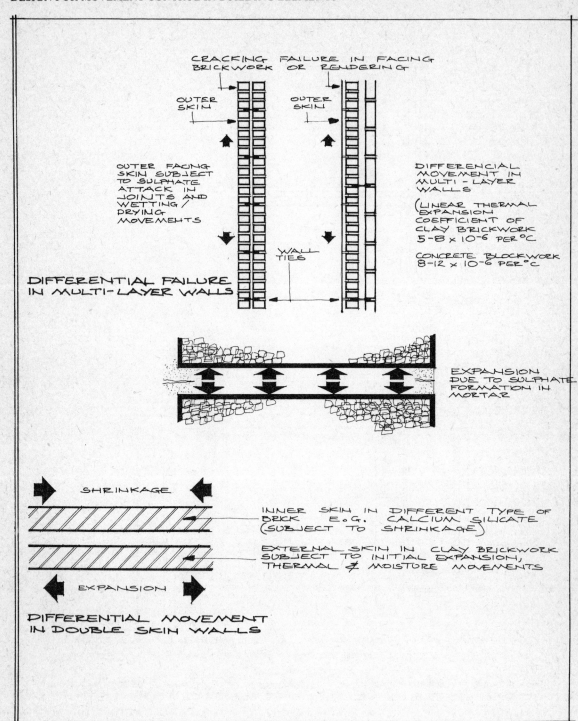

Diagram 5.5.17   Based on diagram page 10 SCP tn 4

WALLS: BRICK AND BLOCK MASONRY

2 Separation from restraining or moving adjoining elements by joints at junctions with adjoining elements (see diagram 5.5.18).
3 Selection of joining materials generally or locally to accommodate movements.
4 Provision of movement joints in the walling.

Division into discrete panels

Discrete panels of walling can be formed by insertion of other components. It is essential that the wall sections are completely separated (see diagram 5.5.19).

As there is no change in the slenderness ratio or loading in the wall, stability is generally not affected by this method.

Separation from adjoining elements

Slip planes and sliding bearings

Slip planes are the simplest form of separation between walls and slabs or beams at the bearing surface. As recommended by BRE Digest 157, the effectiveness of the slip plane is insured only if the top of the wall or bearing is flushed up with mortar and this allowed to set prior to insertion of the slip plane.

Materials for slip planes are:

(a) polythene sheeting (two layers) for every low friction
(b) felt or polymer DPCs for medium friction
(c) metal flashings for medium friction
(d) specialist bearings for high load situations (see diagram 5.5.20).

Load limitations for slip planes

Manufacturers of slip plane materials usually recommend a maximum load of $0.3N/mm^2$ ($4516/m^2$) for slip planes. Higher loads require rubber pads and more elaborate resilient seating mechanisms.

Expansion and contraction joints

Both vertical and horizontal joints are often necessary. At junctions these joints have to allow for the following:

1 Movement in the frame or adjoining element (see diagram 5.5.21–5.5.24).
2 All movements and deformations in the walling.
3 Tolerances: inaccuracies in the frame.
4 Movement accommodation factor of jointing material. See section on joint design.
5 Time of construction, ie weather shrinkage or expansion will be the main factor.

SELECTION OF WALL JOINTING MATERIALS

Mortar mix selection

It is recommended that the WEAKEST mortar is used that can satisfy the walls strength and durability requirements. This helps to accommodate movements by spreading the movement in individual joints by a weaker mix designed to allow movement to take place. This can be used generally in the entire wall (as in traditional walling) or locally at anticipated stress concentrations. Alternatively an open joint can be left and pointed up later if the movement is limited to the early life of the wall (see diagram 5.5.19).

PROVISION OF WALL MOVEMENT JOINTS

Joint requirements: for movement control

Movement joints in the general run of walling are required if the wall length is such as to cause excessive strains due to thermal or moisture movement and to prevent failure due to local stress concentrations (see diagram 5.5.25).

Other required joint performances

External wall joints must maintain the essential functions of the wall, ie they must exclude air, water and dust, be durable and allow for maintenance and repair. In certain cases they must be robust (vandal proof) and sometimes fire resistant, and present a satisfactory appearance.

Any of these factors can be critical in joint type selection and selection of sealant.

DESIGN FOR MOVEMENT CONTROL IN BUILDING ELEMENTS

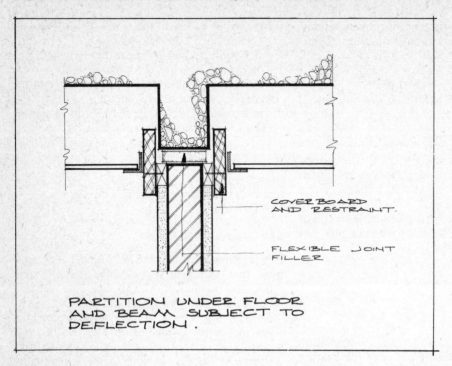

Diagram 5.5.18

Joint profiles: flexible sealed type

Due to the difficulty of forming a labyrinth type of water shed or lapped joint in straight walling thickness, joints are usually sealed by the flexible sealant method (see diagram 5.5.26).

Joint location and selection

Wall joints in long walls may be required *in addition* to the joints at junctions with dissimilar materials and at points of weakness or restraint which must be carefully considered as described earlier.

It is essential to assess as far as practicable the location, direction and extent of all movements likely to occur and assess their effects before finally deciding on the spacing and location of movement joints. Naturally this will also depend on the wall material used, since their movement properties differ greatly (see section 3.4).

Joint spacing

Some typical recommended maximum spacings for various wall materials listed below shows the wide variation in the movement properties of the different walling materials.

Walling material:	Recommended maximum joint spacings (ie panel lengths)
Clay facing brickwork	10–12 m
Calcium silicate brickwork	7.5–0 m
Dense concrete blockwork	7.5 m
Lightweight concrete Blockwork	6 m

Also panels where the length exceeds twice the height

Note These joints need only be the contraction type but if flexibly filled can then cater for a small degree of expansion in individual wall sections.

The correct type of joint to accommodate the anticipated movement needs to be selected and its construction will also depend on other factors, such as the degree of robustness required, which will affect the joint profile, filler and sealant selection.

WALLS: BRICK AND BLOCK MASONRY

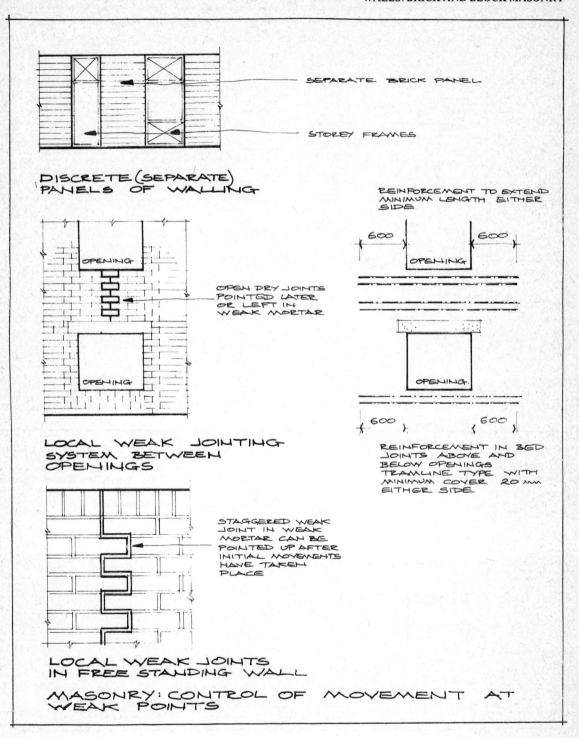

Diagram 5.5.19

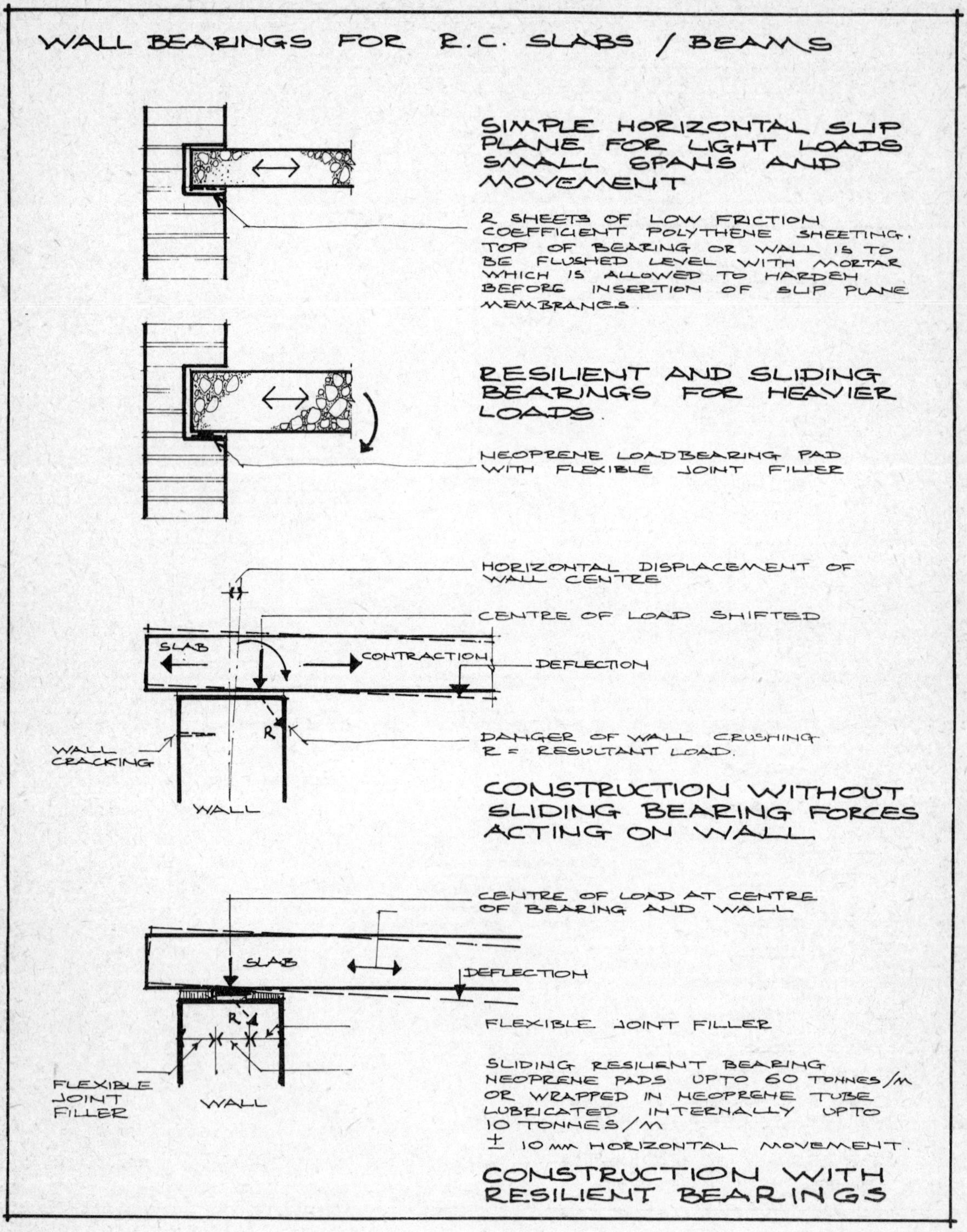
WALL BEARINGS FOR R.C. SLABS / BEAMS
SIMPLE HORIZONTAL SLIP PLANE FOR LIGHT LOADS SMALL SPANS AND MOVEMENT
2 SHEETS OF LOW FRICTION COEFFICIENT POLYTHENE SHEETING. TOP OF BEARING OR WALL IS TO BE FLUSHED LEVEL WITH MORTAR WHICH IS ALLOWED TO HARDEN BEFORE INSERTION OF SLIP PLANE MEMBRANES.
RESILIENT AND SLIDING BEARINGS FOR HEAVIER LOADS.
NEOPRENE LOADBEARING PAD WITH FLEXIBLE JOINT FILLER
HORIZONTAL DISPLACEMENT OF WALL CENTRE
CENTRE OF LOAD SHIFTED
SLAB
CONTRACTION
DEFLECTION
R
WALL CRACKING
DANGER OF WALL CRUSHING R = RESULTANT LOAD.
WALL
CONSTRUCTION WITHOUT SLIDING BEARING FORCES ACTING ON WALL
CENTRE OF LOAD AT CENTRE OF BEARING AND WALL
SLAB
DEFLECTION
R
FLEXIBLE JOINT FILLER
SLIDING RESILIENT BEARING NEOPRENE PADS UPTO 60 TONNES/M OR WRAPPED IN NEOPRENE TUBE LUBRICATED INTERNALLY UPTO 10 TONNES/M
± 10 MM HORIZONTAL MOVEMENT.
FLEXIBLE JOINT FILLER
WALL
CONSTRUCTION WITH RESILIENT BEARINGS

Diagram 5.5.20

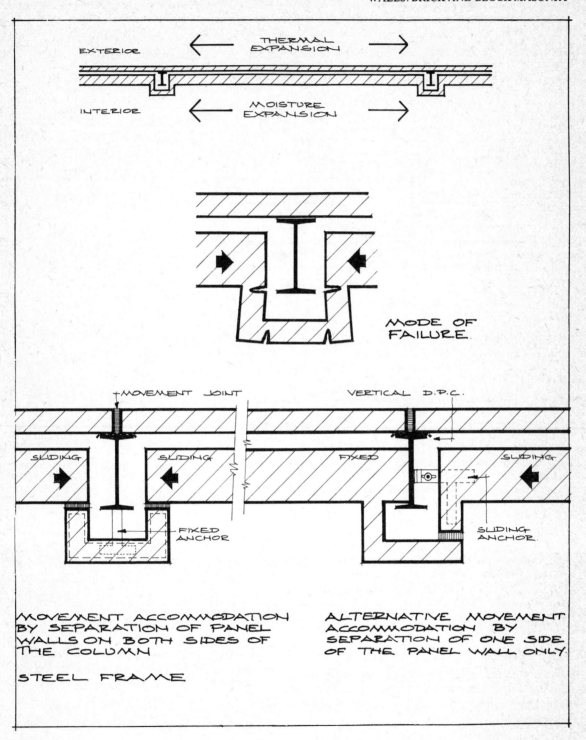

Diagram 5.5.21

DESIGN FOR MOVEMENT CONTROL IN BUILDING ELEMENTS

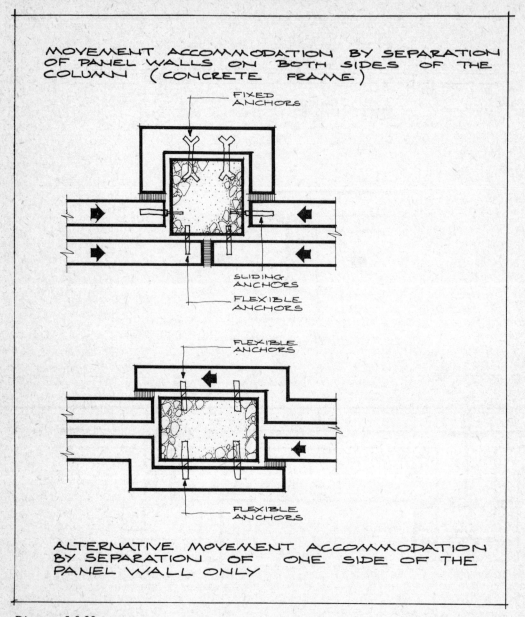

Diagram 5.5.22

Joint arrangement

The arrangement of joints in double skin construction presents the designer with a choice of layout which has to best suit individual circumstances (see diagram 5.5.25).

1 Redlands Bricks Ltd: *Movement joints in clay brickwork.*
2 BRE Digest 157: *Calcium silicate brickwork.*
3 A Tovey C & CA *Concrete Masonry for the Designer.*

SUMMARY OF WALL JOINT TYPES

(see diagram 5.5.26 and 5.5.27)

WALLS: BRICK AND BLOCK MASONRY

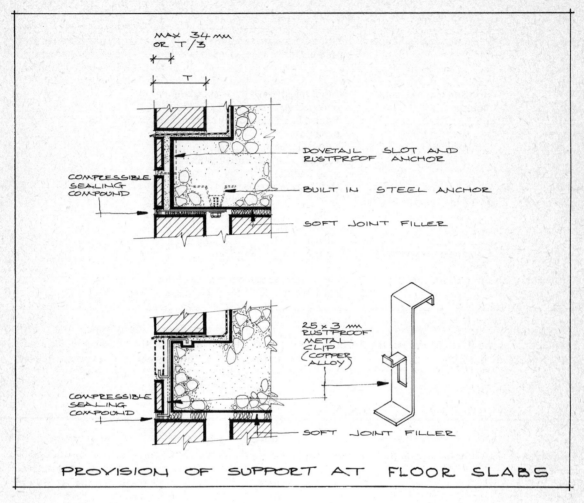

Diagram 5.5.23

1 *Expansion joints*

(see diagram 5.5.27 and table 5.5.6)

2 *Free movement joints for compression or expansion*

These joints require to accommodate movement at right angles to the joint. As this can result in a relatively large change of joint widths, the method of sealing and filling the joint usually require a minimum joint width which will depend on the amount of the movement to be accommodated and flexibility of the materials used for filling and sealing the joint as previously described.

As wall movement is usually greater horizontally than vertically, due to the larger dimensions in this direction these joints are usually vertical.

3 *Contraction joints*

Some wall joints only need to be the contraction type. The joint spacing given earlier applies only to this type. True expansion joints are usually not needed except at 30–50 m centres, provided contraction joints are provided between them.

WALL JOINT FORMATION

Wall joints are formed as the work proceeds either

139

DESIGN FOR MOVEMENT CONTROL IN BUILDING ELEMENTS

Table 5.5.6 (see also diagrams 5.5.19, 26 and 27)

Direction	Vertical movement joints	Within the thickness of the wall suitable for shrinkage or expansion	(from top of wall to DPC only. Unless ground level is over 4 course below)
	Horizontal movement joints	Within thickness of the wall usually only necessary under slabs/beams in framed buildings usually compression or sliding	
Shape	Straight joint	The usual type a straight gap is formed vertically during construction, usually visible unless hidden by a feature for shrinkage or expansion	
	Toothed or staggered joint	Formed to follow bonding not plainly visible	
		Usually for shrinkage only. Can be filled later	
Function	Compression expansion joints	Require flexible filling/seal	
	Contraction only	Require flexible seal but can be mortar filled or left unfilled for internal use	

by building in temporary spacers which are later replaced – or the flexible filler in strip form as the work proceeds (see diagram 5.5.26).

MOVEMENT CONTROL METHODS IN WALLS

Alternative 2: Mitigation/reduction of movement and effects

Prevention: mortar mixes

Prevention measures can reduce the amount of movement in walling and reduce the effects. The use of weaker mortars has already been mentioned as accommodating movements by spreading the effects. Wall movements are mainly due to shrinkage due to moisture loss and differential movements. It is possible to reduce the amount of these movements by correct *design*, *selection* of *materials* and good *workmanship*.

Shrinkage prevention: workmanship

To prevent or reduce moisture loss during construction the lower the moisture content at time of laying the less will be the drying shrinkage.

Temporary movement joints once the initial shrinkage is over may still be advisable. Cover up work which is unfinished and exposed protect blocks from rain prior to laying.

Protection

However protection of masonry, particularly porous concrete blockwork and calcium silicate bricks prior to laying is necessary to reduce shrinkage. It is advisable to allow walls to dry out before applying plaster finishes.

Good bonding

Good bonding will prevent cracking particularly if bonding is poor at stress concentrations, eg windows plus lintel bearings and see diagram 5.5.10.

Selection of wall materials

Selection of suitable materials can do much to reduce the likelihood of movement ie selection of bricks with a low sulphate content, ensuring blockwork has been manufactured for at *least* two weeks, preferably more.

Ensure bricks have been cured.

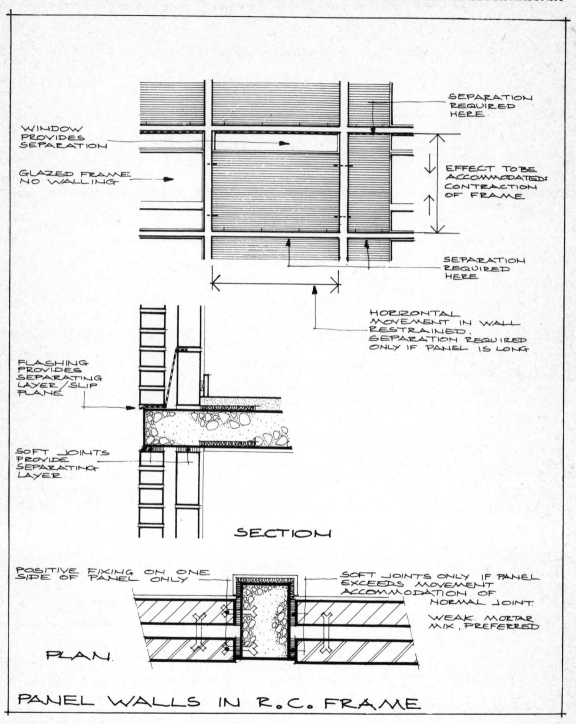

Diagram 5.5.24 (by courtesy of BSCP 121 'Masonry')

DESIGN FOR MOVEMENT CONTROL IN BUILDING ELEMENTS

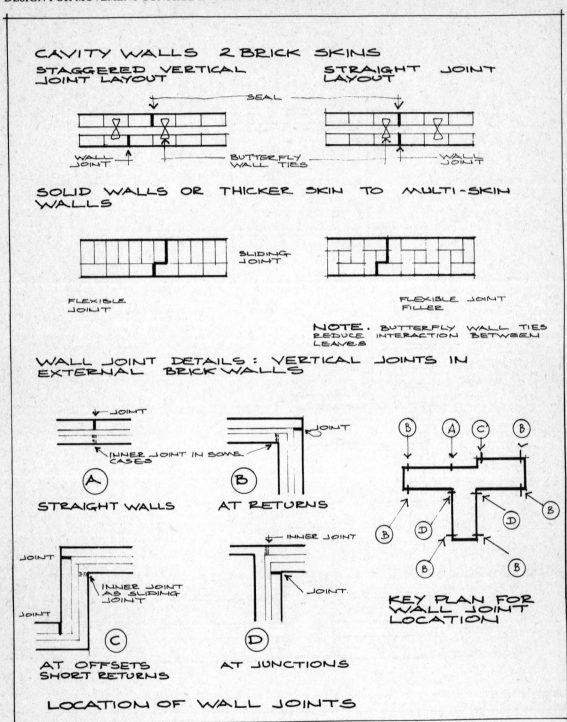

Diagram 5.5.25

WALLS: BRICK AND BLOCK MASONRY

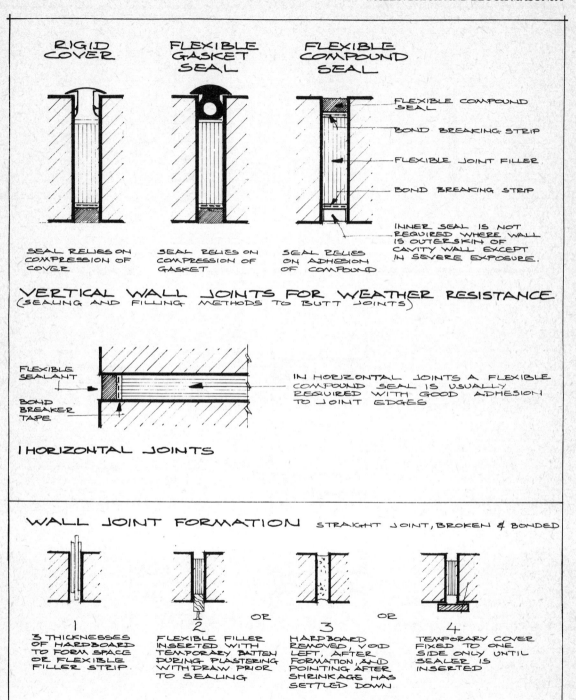

Diagram 5.5.26

DESIGN FOR MOVEMENT CONTROL IN BUILDING ELEMENTS

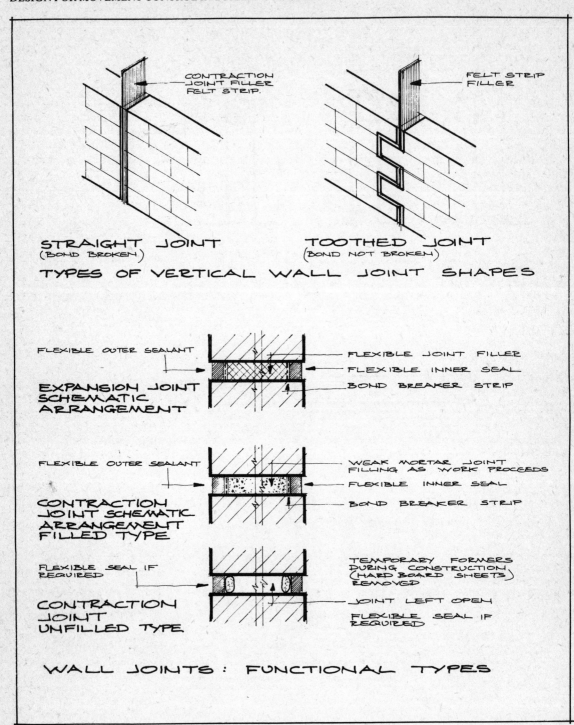

Diagram 5.5.27

Reinforcement: location

This is usually only possible locally to spread out the stress concentration which could cause cracking locally eg above and below window openings where a movement joint may be undesirable (see diagram 5.5.19).

Type

In blockwork, this reinforcement should be the 'tram line' type preferably with an effective diameter between 3–5 mm provided with a cover of at least 20 mm from the face of the mortar and *galvanised* if in an external exposed wall.[1]

5.6 EXTERNAL WALL CLADDINGS AND FACINGS

MOVEMENT IN CLADDINGS AND FACINGS

Definition: continuous or discontinuous support

It may seem useful to regard facings as those finishes or components which are applied directly to a solid or continuous *backing (without a gap)*, eg solid bedded tiling and claddings as components applied to a discontinuous or *intermittant backing/* supports. This can be the structural framework or a subframe supported by the main framework.

Traditional facings

Traditional facings and claddings were usually fairly small units of solid bedded or overlapping rigid units, eg weather boarding or tile hanging. Movement at each joint and differential movement with the backing at each unit and joint was negligible and the movement was distributed.

Larger cladding units, movements and inaccuracies

In cladding supported on intermittant supports which may be 1, 2 or even 3 m apart larger movements both of the supporting structure and the cladding itself will take place and differential or *relative* movement will be more significant, ie as stated in BRE Digest 223 *Wall cladding: designing to minimise defects due to inaccuracies and movements:* 'The problem becomes increasingly important with increase in dimension over which the relative (differential) movement occurs' (see diagram 5.6.1).

The problem of inaccuracies therefore becomes more critical since larger units of facings and claddings will increase the deviations both as a result of inaccuracies in the supporting structure and the cladding itself.

Adequate tolerances to compensate for these deviations inaccuracies must be taken into account in the design of fixings and jointing of these larger cladding/facing units, data is given of the *allowable* deviation in accordance with standard codes of practice. Any larger deviations, although not the designers responsibility, could nevertheless also result in failures.

Corrective measures are known to be very expensive due to scaffolding, replacement of cladding and of failed hidden fixings.

SCOPE

The scope of this section is the identification of causes of movements affecting cladding and the procedures which may be adopted and design precautions to be taken to avoid failures.

Relative movements

In order to reduce the likelihood of failures in cladding due to movements it is 'critically important to examine the potential design for those dimensional changes in service which will produce *relative* movements between cladding and structure or background'* (see diagram 5.6.2).

Design procedure

The procedure suggested by this section is:

1 identification of likely causes and effects assessment of the amount of each contributory movement due to each cause and the combined or discounted effects.
2 assessment of effects at all critical points, eg

---

[1] Tovey A *Concrete Masonry for the Designer*, C & CA.

* BRE Digest 223 'Wall cladding': designing to minimise defects due to inaccuracies and movements.

DESIGN FOR MOVEMENT CONTROL IN BUILDING ELEMENTS

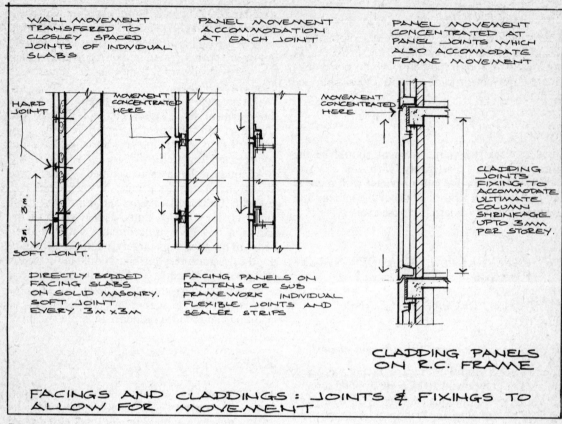

Diagram 5.6.1

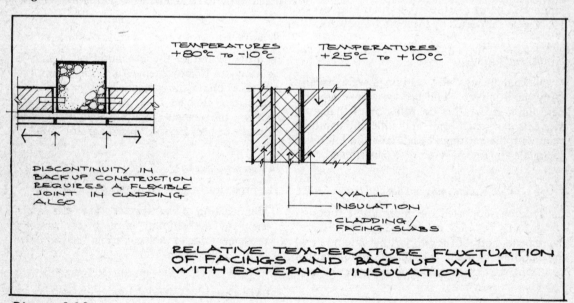

Diagram 5.6.2

bearings, overhangs, stop ends, joints. This depends on the form of the cladding and layout at fixings and supports.
3 choice of suitable details, fixings and jointing and junctions with other components suitable for the selected form of cladding.
4 modification as necessary of the design of the facing/cladding in materials or form.

Joint design is dealt with in detail in a separate section and the choice of joint type and joint location is dealt with as appropriate for various cladding systems.

MOVEMENT CONTROL CRITERIA AND FAILURE LIMITS

Inaccuracies accentuate the failures due to movements affecting cladding:
The main failures as a result of these factors are:

1 Failure of fixings – causing dislodgement or adhesion to backing of facings/claddings.
2 Failure of fixings – or adhesion to bearings.
3 Failure of jointing – causing loss of weather tightness and contributing to failure of fixings.
4 Failure/fracture of the cladding units themselves – causing dislodgement, poor appearance and disintegration and loss of weather tightness and durability.

Any failure of the fixings or bearings is a serious matter affecting safety, since units are often large and heavy or mounted at appreciable heights.

Some cladding materials, eg brickwork can stand considerable distortion before failure in the cladding itself, but distortion of this type is unsightly and can itself lead to further deterioration of the cladding (see diagram 5.6.3).

Since weather tightness is the main criteria of joint failure design of joints to accommodate all anticipated movements without failure of this type is critical.

MOVEMENT CHARACTERISTICS OF FACING MATERIALS

Certain types of movement and effect arise from the movement characteristics and properties of individual materials used as claddings/facings.

EXTERNAL WALL CLADDINGS AND FACINGS

Table 5.6.2, adapted from the table in **BRE** Digest 217 'Wall cladding defects and their diagnosis' is intended as an initial reference to alert the designer to the problems likely to be encountered:

ASSESSMENT OF MOVEMENTS AND EFFECTS

The influence of the form and location of fixings (see diagram 5.6.1)

Relative movements

The effect of the relative movements mentioned earlier are concentrated at the points of contact or fixing between the cladding/facing and the supporting fabric.

Location of effects

If this fabric is continuous as with directly supported facings, then the contact is distributed over the entire area of the facing and the effects will also be similarly distributed.

Types of supporting structure: continuous masonry

Where the facings are directly fixed or bedded, the supporting structure is usually in the form of continuous supporting masonry. However this may support roof or floor slabs and be in the form of panel walls between columns and beams which all break the continuity of the backing and can cause local deformations (see diagrams 5.6.1 and 5.6.2). In continuous masonry the movement described in the section on walling will directly affect the cladding particularly if this is directly attached.

The following summary may also prove useful in assessing the effects:

1 Thermal movement in the backing wall is not as great as in the exposed cladding, particularly with a reflective facing and *external* thermal insulation to the wall.
2 Moisture expansion of brickwork and drying shrinkage of concrete blockwork are likely to be the most serious movements encountered. Note concentrations at stress concentrations: wall failure will usually mean facing failure at these points.

DESIGN FOR MOVEMENT CONTROL IN BUILDING ELEMENTS

Identification of movement failures: causes and effects

The following table is reprinted from BRE Digest 223 *Wall Cladding* referred to earlier.

Table 5.6.1 Movements that may affect cladding

Cause	Effect	Duration, frequency	Examples of materials or components affected	Significance for design
1 Temperature changes	Expansion and contraction	Intermittent, diurnal, seasonal	All. Where restrained, distortion or damage may occur. Distortion may also result from temperature gradients or from non-homogeneity	Extent of movement is influenced by thermal coefficient, exposure, colour, thermal capacity, insulation provided by backing
2 Moisture content changes:				
(a) Initial moisture absorption	Irreversible expansion	Relatively short term, due to absorption of moisture after manufacture	Brick and other ceramic products	Depends on age of product most movement occurs within first 3 months of product's life
(b) Initial moisture release	Irreversible contraction	Relatively short term	Mortar, concrete, sand-lime bricks	May require measures to control or distribute cracking
(c) Alternate absorption release of moisture in service	Expansion and contraction	Periodic – eg seasonal	Most porous building materials, including cement based and wood or wood based products. Restraints, humidity gradients or non-homogeneity may produce distortion. Laminates of dissimilar materials may bow particularly if their construction is asymmetrical	Generally less significant for cladding than are thermal movement but wood experiences large moisture induced movements across the grain.
3 Loading on structure:				
(a) Elastic deformation under service loads	Normally insignificant in vertical members but horizontal members may deflect	Continuous or intermittent under live loads; long term under dead loads	Suspended floor and roof slabs, beams, edge beams or spandrels, of all materials (whether they support or 'contain' the cladding)	Needs consideration in relation to fixings and bearings for cladding and to possible compression of 'contained' cladding: deflections in pre-stressed concrete members may be relatively large
Creep	Contraction of vertical and deflection of horizontal members	Long term	Reinforced and pre-stressed concrete components as above	Needs consideration as above. May also be significant where load bearing concrete walls or columns have cladding such as mosaic or other tiling directly bonded

EXTERNAL WALL CLADDINGS AND FACINGS

continued

Cause	Effect	Duration, frequency	Examples of materials or components affected	Significance for design
4 Wind loading on cladding	Deflection	Intermittent	Lightweight cladding, including fixed and opening glazing; sheet siding	Extent of deflection depends on exposure for a given stiffness. Deflection is commonly designed not to exceed 1/240 of the span in order to avoid damage to sealants or glazing
5 Chemical changes: (a) Corrosion	Expansion	Continuous	Iron and other ferrous metals	Depends on protection or on corrosion resistance of material; electrolytic corrosion may require consideration. Corroding fixings can seriously disrupt cladding
(b) Sulphate attack	Expansion	Continuous	Portland cement based products in construction where soluble sulphate salts (eg from high-sulphate bricks) and persistent dampness present	Significant for cladding where the construction affected has cladding such as mosaic or other tiling or rendering, bonded directly to it.
(c) Carbonation	Contraction	Continuous	Porous Portland cement: products, such as concrete, lightweight concrete, asbestos-cement	Not very significant unless distortion might result – for example, asbestos cement sheets painted on one face only
6 Vibration (from traffic, machinery wind forces)	Generation of noise, possible loosening of fixings, disturbance of glazing seals	—	Lightweight cladding, sheet siding	Noise discomfort to occupants; possible rain penetration past seals by 'pumping' action of glazing or spandrel panels. Natural frequency of cladding or panels may influence response
7 Physical changes: (a) Loss of volatiles	Contraction, loss of plasticity	Short or long term depending on materials, exposure	Some sealants, some plastics	Contributes to age-hardening of some sealants. May lead to embrittlement and distortion of some plastics.
(b) Ice or crystalline salt formation	Expansion and possibly disruption in some building materials	Dependent on weather conditions	Porous natural stones, very exposed brickwork	Damage usually confined to spalling and erosion of surfaces

DESIGN FOR MOVEMENT CONTROL IN BUILDING ELEMENTS

Table 5.6.2 Some common causes and effects of movement

Cladding/facing material	Type and causes of movement	Effects of movement
Fairfaced brickwork, rendered brickwork, brickwork with directly fixed tile facing	Sulphate action on mortar causing expansion	Ion of mortar, durability strength, restraint by frame, bowing and horizontal cracking.
	Initial drying, shrinkage of finish	Detachment of finishes, general fine cracking of finish.
Brick slips overflood/RC frame	Moisture expansion of brickwork, shrinkage of frame	Cracking at floor levels and dislodgement of brick slips
Stone or precast concrete cladding	Corrosion of fixings, expansion	Corrosion stains on surface, failure of fixings and support
	Differential movement between frame and cladding, thermal movement in cladding. Creep or other deformations in structural frame.	Cracking of cladding, spalling at edges and corners. Misalignment
Light cladding panels in frames or frameless panels	Thermal movements in panels and supporting framework, differential movement with relatively shuttered structure or supporting wall	
Glazed framed spandril or curtain walls	Thermal movement of framework. Concentrated at mullion/transom junctions wind load deformation of glass	Rainwater penetration, distortion of retaining beads, detachment of sealant from glass

3 Settlement in foundations: obviously such movement will be directly transferred.
4 At points of discontinuity where backing material changes (see diagram 5.6.1) movements in the backing may be concentrated and affect the cladding locally.
5 When fixed over *in situ* concrete walls, drying shrinkage and creep in the concrete.

Types of supporting structure: cladding supported by skeleton framed structures and on intermittent supports.

Effect of deformations in the structural framework

Both steel framed and reinforced concrete framed buildings are subject to deformations in the frame which can affect the cladding.

As described in the section on frameworks, the deformations in frameworks can affect individual members of the frame, the junctions and the position of these junctions due to overall movements of the frame (see diagrams in section on frames.)

However, these will only affect the cladding if the relative dimensions to the cladding undergo a change, such as overall shortening of the framework due to creep in columns, deflection of slabs if supporting the cladding, (see diagram 5.6.3).

Generally deformations in steel framed buildings tend to be larger initially as the loading is applied during erection. Reinforced concrete buildings being generally stiffer tend to move less initially but the time related creep and shrinkage effects can be significant.

LOCATION AND DIRECTION OF DEFORMATIONS DUE TO MOVEMENTS IN THE SUPPORTING STRUCTURE

Concentrated at supports

Relative movements between the frame and the cladding are transferred at the supports or fixings of the cladding, from here the induced stresses and

EXTERNAL WALL CLADDINGS AND FACINGS

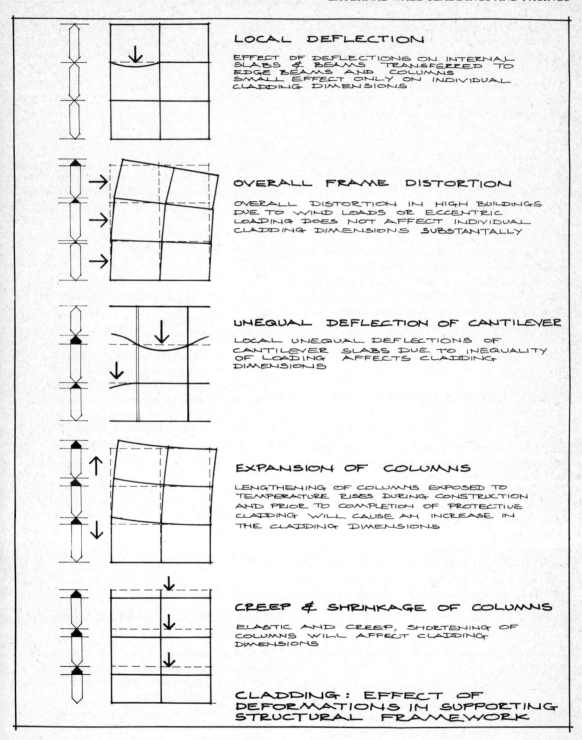

Diagram 5.6.3

DESIGN FOR MOVEMENT CONTROL IN BUILDING ELEMENTS

movements are distributed into the cladding (see diagram 5.6.4A).

Since the movements of the structure and cladding can be a combination of linear distortion and bowing, ideally provision at fixings should be multi-directional but certainly in the three major axes, ie the points of attachment and fixings require allowance for displacement due to both movements and inaccuracies.

However in practice it may only be necessary to allow for movements in predetermined limited directions (see diagram 5.6.4B).

Influence of fixing details and type

The fixings detail as well as their location can influence the magnitude and direction of movements in the cladding as is shown in diagram 5.6.5 adapted from BRE Digest 223 wall cladding designing to minimise defects due to inaccuracies and movements.

ASSESSMENT OF MOVEMENTS AND DEFORMATION IN CLADDINGS/FACINGS

In order to determine the relative movements between supporting fabric and the cladding both movements in the supporting fabric and the cladding itself must be estimated.

Both the type and location of movements have been described earlier. The actual estimation process is described in the section on estimating movements. The combined effects of all movements needs to be determined and then added or subtracted as appropriate to obtain the overall combined relative movement as is shown by the following example.

Cladding: estimation of combined relative movements

(a) *instantaneous deflections*: subsequent to early fixing of cladding can be substantial.
(b) *long term deflection creep and shrinkage in same direction in concrete structures only*
    shrinkage say 0.05% or 500 microstrain
    creep say   0.04% or 400 microstrain

(c) thermal deformation see materials section, eg
dense concrete panel 3 m high for temperature
rise of 50°C   1.8 mm
   Brick panel   0.9 m
   Lightweight block   1.1 mm
   GRP approximately   4 + 5 mm

Additive: relative effects

Since expansion of the cladding is opposite to the compression in the opposite direction the relative effect is the total combined quantity of all these movements, eg in a 3 m storey height RC building:

1	Deflection (instant)	1.0 mm
2	Shrinkage	1.5 mm
3	Creep	1.2 m
	Total	3.7 mm

add for differential relative movements:

4	Thermal in cladding	1.8 m
	Total movement	5.5 mm

This is a vertical movement per storey height and could be concentrated at one fixing and joint.

If sheet steel faced or GRP cladding panels had been used in this example with a much higher panel thermal movement the relative or combined movement to be allowed for at joints and fixings would have been correspondingly greater.

DETAILED DESIGN TO ALLOW FOR MOVEMENT

Directly fixed facings

Jointless facings: rendering

The accommodation of movement in renderings is achieved mainly by

1 *mix design* using as weak a mix as possible, sharp sand and preferably lime additive.
2 provision of movement joints to break up large areas (see diagram 5.6.6).
3 good adhesion by choice of suitable background material or treatment to overcome either excessive suction or low suction backgrounds.

EXTERNAL WALL CLADDINGS AND FACINGS

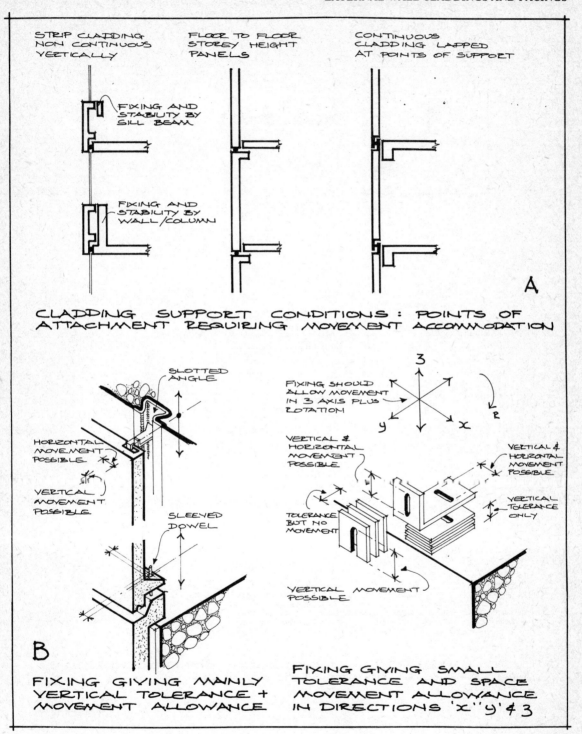

Diagram 5.6.4

DESIGN FOR MOVEMENT CONTROL IN BUILDING ELEMENTS

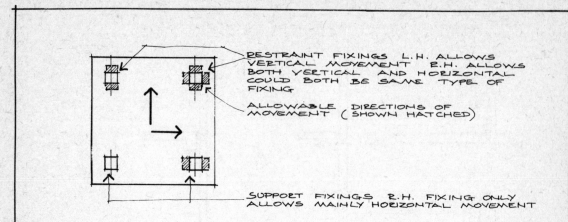

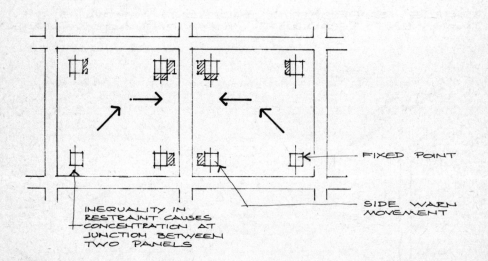

MOVEMENT IN CLADDING CONTROLLED BY FIXING TYPE LOCATION

Diagram 5.6.5

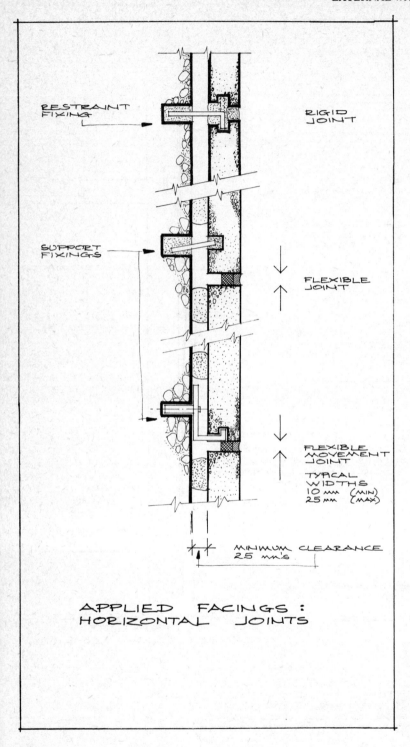

Diagram 5.6.6

DESIGN FOR MOVEMENT CONTROL IN BUILDING ELEMENTS

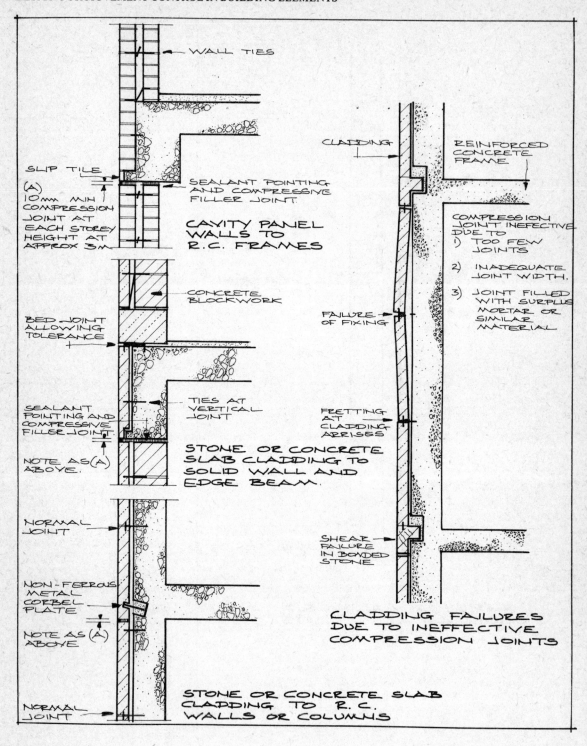

Diagram 5.6.7

EXTERNAL WALL CLADDINGS AND FACINGS

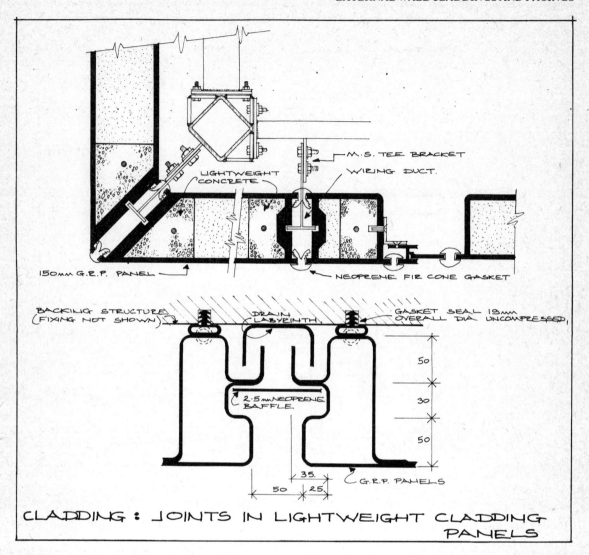

Diagram 5.6.8

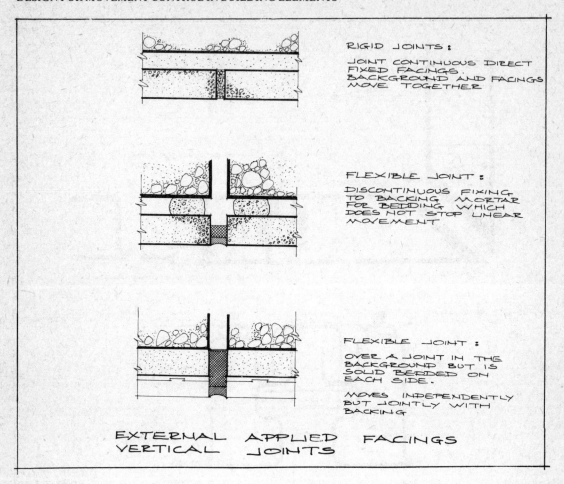

Diagram 5.6.9

Jointed solid bedded facings

eg, stone or tiles movement is distributed over the entire area of the facings, but never the less full flexible movement joints should be provided at least every 3 m in both directions to allow relative movements (see diagrams 5.6.6 and 5.6.7).

Claddings with intermittent fixings to structure framework: columns beams or floor slabs

Points of support

As has been pointed out the location and type of fixing can affect the directions and the location and magnitude at certain points and joints of the relative movements, especially when cladding is fixed directly to frames, ie without mullions or subframes.

Inaccuracies and tolerances

Both the jointing and fixings must allow for considerable inaccuracies which have the effect of displacement of the fixings or the joint dimensions *before* the movements take place, ie the overall effect of inaccuracies and movements on fixings and jointing is *additive*.

Accuracy criteria

Accuracies have been investigated and deviations from the mean or 'design' dimensions are expressed as maximum allowable deviations from the mean in BS 5606 1978.

The following is an extract:

Permissible deviations for in situ concrete	
Space between walls at flow	± 30 mm
at soffit	± 30 mm
Space between columns at floor	± 20 mm
at soffit	± 25 mm
Structural floor level flatness	± 20 mm
variation in datum	± 20 mm
Precast with *in situ* topping flatness	± 20 mm

It is clear therefore that both fixings and cladding joints must allow for considerable deviations as well as movement deformation.

Diagrams 5.6.4B and 5.6.8 and 5.6.9 show some typical joints and fixings which allow for these.

ROOFS AND ROOF FINISHES

SCOPE

Roof structures and finishes are both covered by this section, in view of the close interaction of movements in all parts of the roof construction, ie the various layers, all of which contribute to the total performance of the roof system.

DESIGN CRITERIA

The main criteria for movement control in roofs are:

1. Integrity and durability of the weatherproofing. Flat roof finishes are the most vulnerable due to low movement tolerance.
2. Integrity of the roof structure.
3. Interaction with supporting or adjoining elements of frameworks.
4. Integrity of attached ceiling finishes if any and integrity of the supporting structures.

Failure limits

The failure limits of roof construction are therefore determined by the effect on the finishes or weatherproofing membrane and secondly by any detrimental effect on the supporting or adjoining structures as well as on the roof structure itself.

As in other elements the amount of movement which can be accommodated without damage of failure depends not only on the strength and durability of the affected components, but also on the manner of their assembly. This can greatly affect the stress induced by movements and can reduce the effect of movements being transmitted between layers, particularly of roof finishes.

The general mechanical behaviour of flat bituminous roofing systems has been fully described by Bonafont in the two papers given as references.

Mechanical properties of waterproof roof covering

The resistance to movements of various types of roof coverings varies greatly depending on their form, material and assembly.

Pitched roofs

Pitched roof coverings being the overlap unit type can stand considerable movements without failure or causing loss of weathertightness.

Flat roofs

Flat roof coverings on the other hand being continuous, rely on their complete integrity from cracks to maintain weather resistance.

Effect of layers

The mechanical strengths of the top (water excluding) layers of multi-layer coverings will depend on the amount of stress transmitted by the base fabric and the strength of this in turn, in the case of bituminous impregnated-coated fabrics, is the strength of the fabric itself.

Most traditional roofing felts suffer from the weakening effects of repeated loadings. Sudden movements are accommodated by elasticity and longer term movement cycles can also be accommodated by a certain amount of viscous flow. The stresses in the top layer also depends on the amount of stress relaxation which can take place between the layers in the time, ie the rate of straining effect. It also depends on the distance from the substructure or cause of movement (see diagram 5.7.1) and on the amount of slip due to partial bonding. Single layer roofing systems are

DESIGN FOR MOVEMENT CONTROL IN BUILDING ELEMENTS

naturally less able to reduce strains than partially bonded layers.

Critical points or points of weakness

These can, as with other elements require special attention and may affect the limits of movement control in the rest of the roof and supporting construction. These are dealt with in more detail later (see diagrams 5.7.2 and 5.7.3)

FAILURE LIMITS

The weatherproof membrane and base fabric

Failure in the function of the weatherproof membrane can take the following forms:
1 Failure to discharge rainwater by ponding (see diagram 5.7.2).
2 Local movement blistering, ripples cockling.
3 Slippage or creep.
4 Rupture: Local cracking or holes.
5 Delamination or uplift.
6 Puncturing.
7 Degradation of surface material.

(1) to (5) can be brought about or be aggravated by movements in the substrate or structure or in the weatherproofing layer itself.

The place of movement in the failure of flat roof finishes can be summarized as follows:

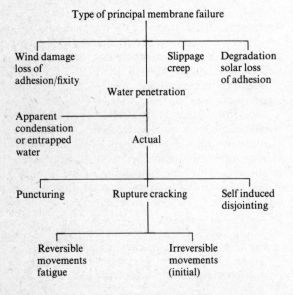

Failures in the waterproof membrane:

See diagram 5.7.3

1 Abutments/expansion joints.
2 Breaks due to upstrands/penetrating components, eg roof lights. roof structures/supports.
3 Edges.
4 Changes of slope or plan area.
5 Points of local changes in substrate.
6 Joints or line of bearings below substrates.

At these points local or increased stresses can be imparted to the covering membrane by differential or locally greater movements in the supporting substance or adjoining elements. Special care is needed in laying of the roof covering at these points and minor movement joints may also be required as detailed later.

Effects on adjoining elements

Since roofs are exposed to the greatest solar heat grains ($+80°C - 10°C$) and the most rapid cooling they are, after foundations, the most critical element to be considered in the total building. Movement in the roof will affect all the supporting structures, either walls or structural frameworks. See section on frameworks for further details and see diagram 5.7.2.

In the case of wall supports an overall angular distortion of 1:2500 would be acceptable. So taking a 6 mm high wall or *maximum* horizontal movement of 2.4 mm would be acceptable, unless a complete sliding roof bearing can be assured. This would limit the roof movement allowable.

Ceiling finishes

The same deflection limits apply to roof slabs as floor slabs, ie 1/300 being maximum allowed to prevent cracking of plaster finishes directly applied to the roof structure.

CAUSES AND EFFECTS OF MOVEMENT IN ROOFS

As shown in the previous section on failure limits, the main effects of interest to the designer are:

1 Failure of roof waterproof finishes due to movements in the substructure.

ROOFS AND ROOF FINISHES

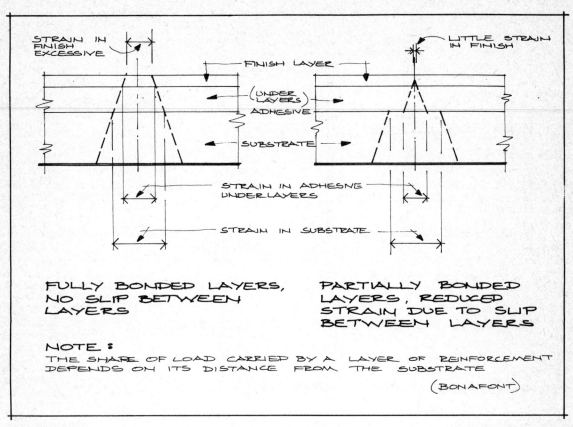

Diagram 5.7.1

2 Failure of the supporting/adjoining elements.
3 Differential movements.

The causes of movements which can cause these failures are shown in Table 5.6.3:

Table 5.6.3

Cause	Effect
Extrinsic causes	
Loading: dead live loads and wind	Excessive deflection of main roof structure. Initial and creep. Excessive uplift over supports
Settlement of supports	Movement in supporting structure or foundations
Differential movements Unequal span supports	Deflection, hogging uplift

continued

Thermal effects	Expansion, differential movement contraction mostly linear but also local buckling, curling, blistering, loss of adhesion
Intrinsic causes	
Initial shrinkage of concrete/screeds	Contraction cracks and splitting
Irreversible movement	
Initial drying out of timber or decking/substrata	
Reversible movement	
Moisture movement in deck and structure	Timber and decking material

161

DESIGN FOR MOVEMENT CONTROL IN BUILDING ELEMENTS

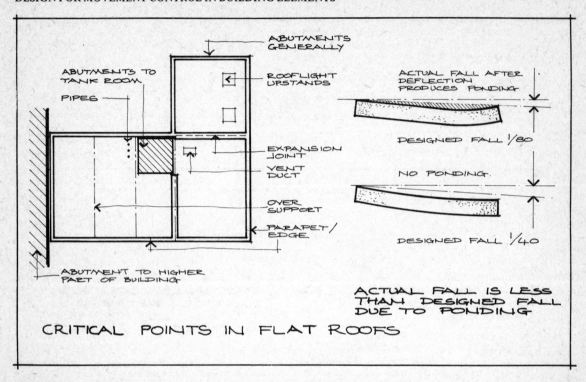

Diagram 5.7.2 Based on PSA *Flat Roofs* Technical Guide

Time dependent movements

The following different timing for movements also needs to be identified, as the effect on various parts of the roofing system varies.

Initial movements At or after installation, eg concrete drying out

Operating movements Temperature ($-20°$C to $+80°$C)

Seasonal as well as diurnal changes produce variations in the operational movements of the roof. Effects will also depend on the time when the roof was erected. Sudden movements, sudden changes of temperature can have detrimental effects on the coverings, since these are usually more capable resisting gradual movements.

These can be in the order of $\pm 60°$C. 50°C changes have been recorded in 20 minutes.

Repeated movements

Also, repeated movements can cause fatigue in the finish as mentioned previously.

MOVEMENT CONTROL METHODS AND PROCEDURE

Assessment

The first step is to identify the causes of movement in the particular circumstances and then to evaluate the extent of the movement in the roof structure which is to be accommodated.

Leaving aside the effect of major subsidence in the supporting structure, for which special joints would be required, the main causes usually are thermal and moisture effects and are either initial and irreversible or normal reversible in operation.

Unrestrained movements

Finally, to analyse the extent to which this movement will be transmitted to the finishes:

As described in the section on movement estimation, data is available on the temperature

ROOFS AND ROOF FINISHES

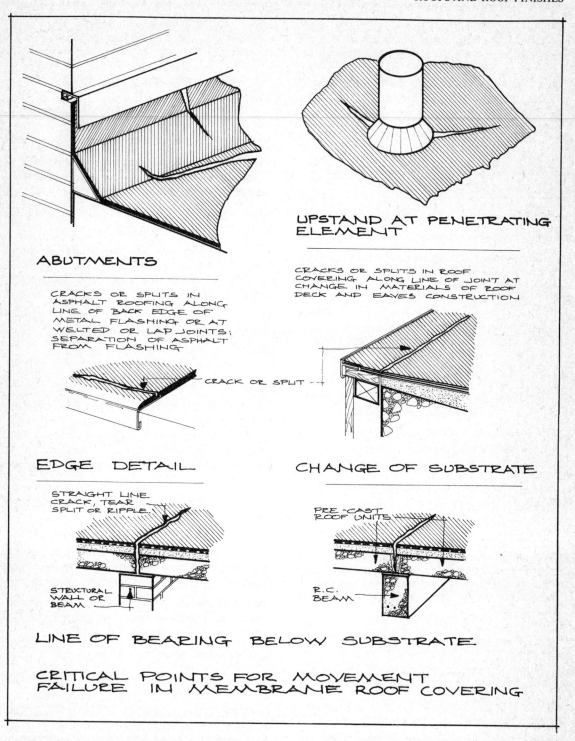

Diagram 5.7.3 Based on PSA *Flat Roofs* Technical Guide

DESIGN FOR MOVEMENT CONTROL IN BUILDING ELEMENTS

ranges and moisture contents in service. From these and the movement characteristics of the materials it is possible to estimate the unrestrained movements anticipated. The degree of restraint varies with many factors, but *half* the unrestrained movement is a useful approximation for design purposes, ie of the actual restrained movements in practice.

Example calculations for movements in roofs
Movement in supporting structure

Initial (irreversible) movements

a *Concrete roof slab:* Assume without reinforcement for *unrestrained* movement calculation. Length 25 m assume total shrinkage at 0.5 mm/m

Free moisture
Movement = range of mean moisture content
× material movement coefficient
× size of member
$M = -0.5$ mm/m × 25
$M = -12.5$ mm
(say ½ restrained = *6.25 mm*)

b *Timber flat roof joists:* Redwood joists 250 mm deep installed at 21% moisture content reducing to annual average of 10% in service.

To calculate traverse shrinkage (across depth)

$M = (21-10) \times -2.8 \times 0.225 = -6.9$ mm change in *depth.*

Note This could be less if drier when installed and kept dry during construction. This contraction is sufficient to cause crack failure in an *upstand* of roof finish if this is rigidly attached to a rigid abutment. However, the upstands may not be formed before this movement has taken place.

Normal operating movements (reversible)

a *Reinforced concrete slab* Assume 25 m long *mean* temperature variation 58°C hot summer to 0°C winter.

Free thermal
movement = Range of mean temperature
(max − min 0°C)
× Material expansion co-efficient per 1°C temperature change ($\times 10^{-6}$)
× Size of member mm
$T = (58-0) \times 0.01 + 25$ (or $\times 10 \times (10^{-6}) \times 25$)
$T = 14.5$ mm variation in length (disposed about the *average* at time of installation)

c *Timber: redwood joists* (in ventilated roof) 225 deep as before range of moisture content 18% summer to 18% winter.

$M = (18-10) \times 2.8 \times 0.225$
$M = 5$ mm variation in depth.

Note This is disposed about the average depth after initial shrinkage.

For data on timber and concrete shrinkage see section 3.

MOVEMENT IN SUBSTRATA/ROOF-DECKS

Cement/sand screeds

Screeded substrates of cement/sand are subject to drying shrinkage of 0.05%, with a greater shrinkage for lightweight aggregate screed. Wet cement screeds should be laid in areas not > 9 mm², checkboard or strip to minimise overall shrinkage.

Plywood chipboard deck

Both are subject to initial drying shrinkage and thereafter normal reversible wetting and drying movement in service as shown in the following example:

Chipboard decking 19 mm deep, felt covered over ventilated space moisture range 8% summer − 18% winter, boards 1.8 m long.

for longitudinal moisture movement

$M = (18-8) \times 0.3 \times 1.8$ m
$M = 5.4$ mm variation in length.

Note This is unrestrained. Good nailing may reduce this to say 3 mm with say 1.5 mm at each joint. This can be accommodated by good quality finishes with partial bonding.

ROOFS AND ROOF FINISHES

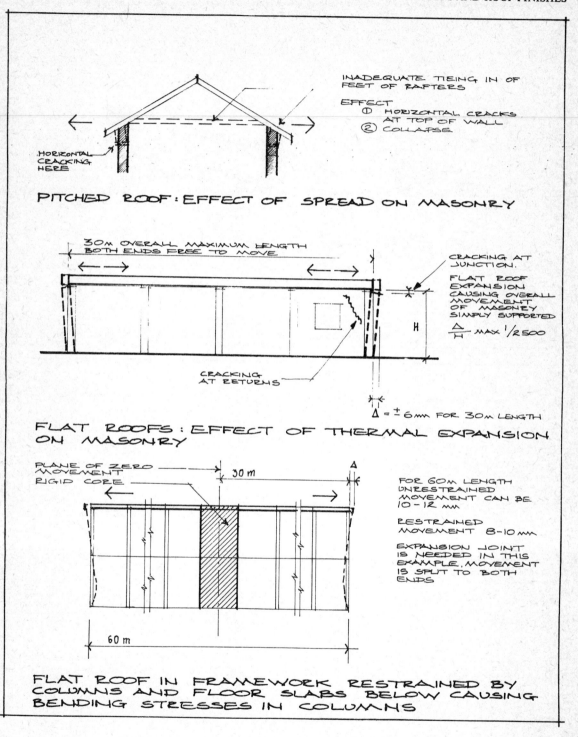

Diagram 5.7.4

DESIGN FOR MOVEMENT CONTROL IN BUILDING ELEMENTS

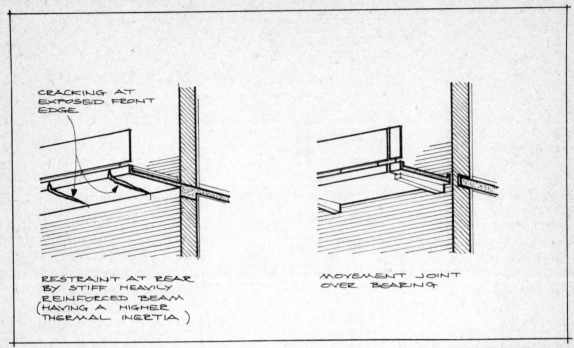

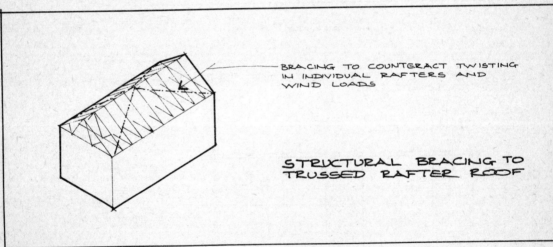

Diagram 5.7.5
Movement in trim and associated components

The following worked example illustrates the extent of the problem:

Aluminium coping/ (edge trim would be similar)
2.3 m length temperature range of hot sun at 68°C to −5°C (winter)
T = (68 − (−5)) × 0.025 × 2.3 m.
T = 5 mm.

Note Over 2.3 m this is a substantial percentage movement. However with screwed fixings at 300 centres in the case of Aluminium edge trim the movement might be restrained to say half this figure, giving 4 ÷ 2 ie 2 mm at each joint. This could be acceptable with a good bedding compound.

MOVEMENT CONTROL METHODS IN ROOFS

Pitched roofs: structure

Structural movement due to 'spreading' of roof structures is the main cause of failure. Warping and twisting of individual roof trusses can cause serious structural failures. These movements are usually controlled by adequate structural bracing (see diagram 5.7.6 (a)).

Junctions

Junctions with abutments can still cause movement problems, but these are usually caused by movement in the abutment, eg moisture movement of brickwork over a flashing/DPC weakened layer and require no precautions in the roof itself.

Flat roofs

These are seriously affected by movements due to the effects of solar gain, radiation loss as discussed earlier.

Since all components form part of a layer system, as shown in diagram 5.7.6 (a–c) all movements are to a greater or lesser extent transmitted from one layer to the other.

Substrata movement control

It is therefore advisable to insert an isolating membrane between the waterproof membrane and substrate unless adequately loaded. This can be *partially* fixed with dots to allow some movement without danger of uplift due to wind.

The actual movement in the substrate can be reduced by correct mix design and laying of screeds and adequate fixings with staggered joists of sheeted decking (see diagram 5.7.7).

Minor movement joints

At junctions in the substrate particularly at changes of material, minor movement joints should be provided as shown in diagram 5.7.3.

Major movement joints

At 30 m spacing in reinforced concrete roofs, or at points of stress concentration as shown in diagram 5.7.2 major movement joints should be provided. These should if possible coincide with wall joints as shown in diagram 5.7.7 and be constructed wherever possible as a raised kerb (see diagram 5.7.8 and 5.7.9).

If traffic is unavoidable (car decks) more sophisticated flush joints will be needed.

The junction between a major joint and a wall joint needs special care. A detail recommended by the GLC/ILEA handbook would be suitable for use in the balcony joint mentioned earlier in this section.

Movement control at supports, abutments/junctions/rood edges

These are the points where either reverse bending, differentials in deflection and differential movements between elements have their effect on the roof finish and are a common cause of failure in bituminous or other jointless materials as shown in diagrams 5.7.10, 5.7.13 and 5.7.14 a minor movement joint is advisable.

Major movement joint at abutments

Where a comparatively flexible deck or roof in timber abuts a brick/parapet an upstand independent of the abutment forms a major movement joint (see diagram 5.7.11).

Where a concrete roof has a major (settlement) joint at a wall abutment a detail such as that shown in diagram 5.7.12 allows for adequate vertical and horizontal movement.

Movement joints in jointed sheeting

Flat metal roof sheets have incorporated upstand and welted movement joints in traditional sheeted roofs for many years and are well known.

Recent developments in new materials, flat stainless steel sheet and aluminium flat sheet adopt similar principles, as do pre-bonded sheets.

Profiled roof sheeting may also require major movement joints where coinciding with a settlement joint.

Details are based on BS CP DC 11822.

DESIGN FOR MOVEMENT CONTROL IN BUILDING ELEMENTS

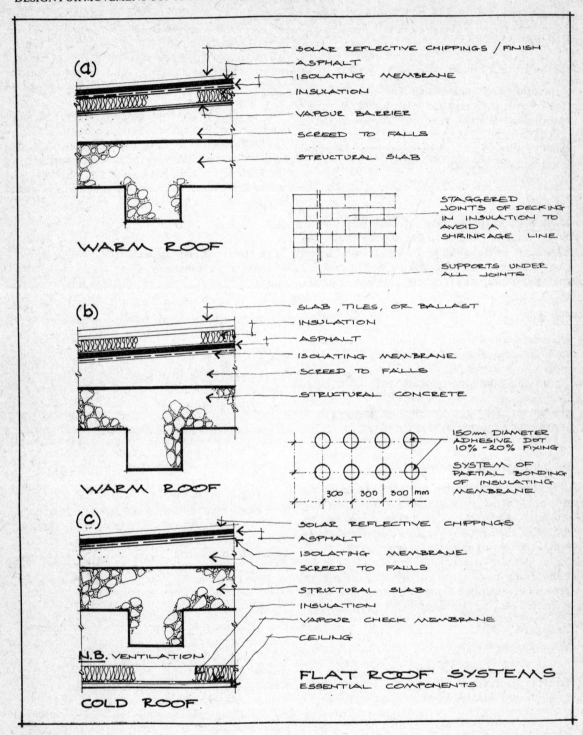

Diagram 5.7.6

INTERNAL SUSPENDED FLOORS

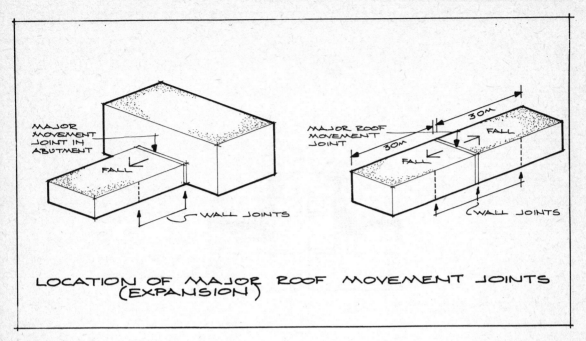

Diagram 5.7.7

5.8 INTERNAL SUSPENDED FLOORS

SCOPE

This section deals with structural floor elements and associated finishes and components in outline only. A detailed treatment of movement problems and control of finishes is given in the section on floor and ceiling finishes as these may require movement joints without affecting the structure.

CLASSIFICATION OF FLOOR TYPES

For the purpose of evaluating the movement characteristics of suspended floor constructions certain broad classification groups are useful:

1. Simply supported floors on loadbearing wall supports: These can be timber or beamed precast or *in situ* concrete slabs. It is the lack of continuity which gives the movement characteristics of this type
2. Simply supported floors on a steel or concrete frame (where no substantial continuity is achieved at the junctions with the frame).

Most composite structures are of this type. The movement characteristics are similar to type 1.
3. Continuous *in situ* concrete floor slabs forming part of a continuous RC column and beam framework.

DESIGN CRITERIA FOR FLOORS

Failure limits

Failure limits in suspended floors are usually the integrity of the finishes. However, failures in the partitions supported on the floors or supporting the floors (see diagram 5.8.1) are a critical factor and see section 5.5 *Walls*, for failure limits of partitions supported on floors where the angular deflection is given as a measure of the deflection limits of partitions in general.

It should be emphasised that local weak points in say a partition or a floor finish could cause cracking or detachment below these general deflection limits, ie for the general condition which formed the basis of the floor design (see diagram 5.8.1). As will be shown, local and differential effects can be serious. Each case must therefore be

DESIGN FOR MOVEMENT CONTROL IN BUILDING ELEMENTS

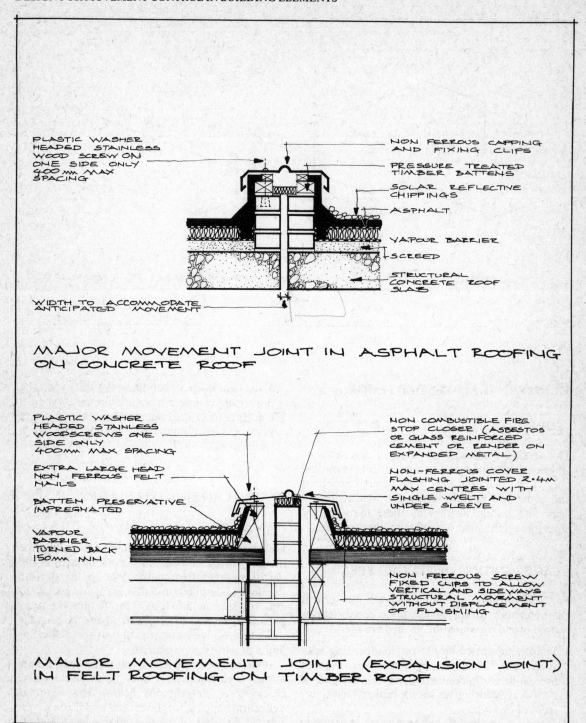

Diagram 5.7.8 Based on PSA *Flat Roofs* Technical Guide

INTERNAL SUSPENDED FLOORS

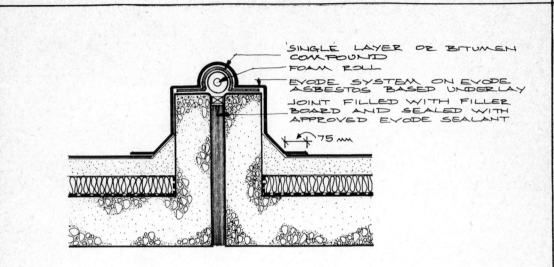

ALTERNATIVE DETAIL AT A MOVEMENT JOINT

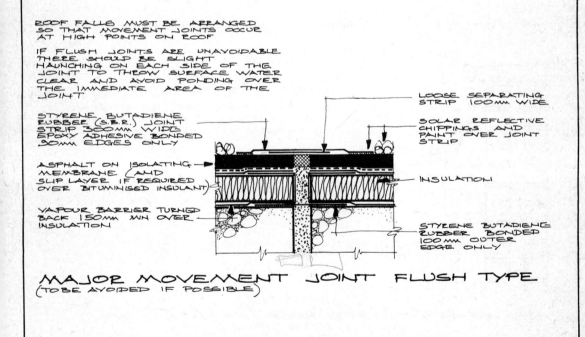

MAJOR MOVEMENT JOINT FLUSH TYPE
(TO BE AVOIDED IF POSSIBLE)

Diagram 5.7.9   Based on PSA *Flat Roofs* Technical Guide

DESIGN FOR MOVEMENT CONTROL IN BUILDING ELEMENTS

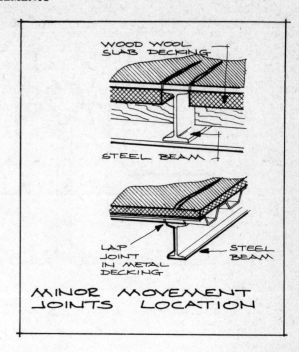

Diagram 5.7.10

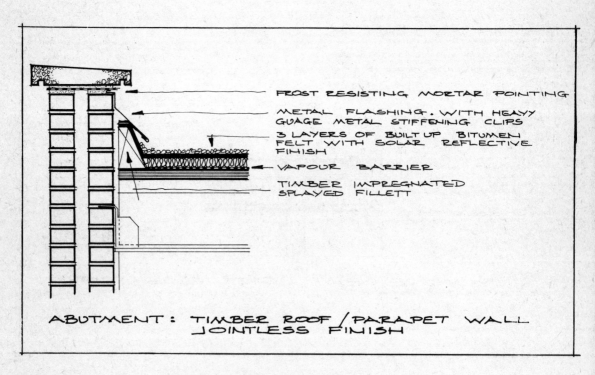

Diagram 5.7.11 Based on PSA *Flat Roofs* Technical Guide

INTERNAL SUSPENDED FLOORS

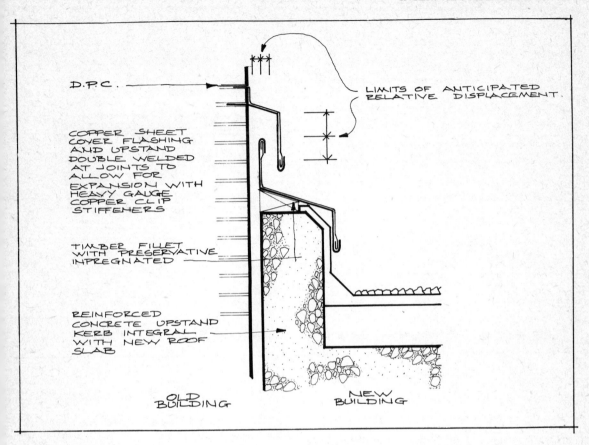

Diagram 5.7.12

critically examined by the design team for possible local effects.

As shown later, the effects of movements in floors are not limited to deflection of the floor itself. However as this is the most common source of deformation, failure limits are usually exposed on this basis as follows:

Deflection limits: timber floors

CEP 112 Part 2 1971: *The Structural use of timber*. The code states that: 'The dimensions of the flexural members should be such as to restrict deflection within limits appropriate to the type of structure, having particular regard to the possibility of damage to surfacing materials, ceilings, partitions and finishings and to functional needs generally in addition to aesthetic requirements': For floors this recommendation may be assumed to be satisfied if the deflection of the supporting members, *when fully loaded* does not exceed *0.003 of the span.*

Deflection limits: reinforced concrete floors

Since precast and particularly prestressed floor beams and floor units are often subject to larger deflections due to absence of continuity and prestressing itself, the maximum deflection limits or adequate stiffness have to be precisely controlled by observing the maximum values of span/depth ratio given in CP 116 1969 *The Structural use of precast concrete* clause 314 *Stiffness of members*.

173

DESIGN FOR MOVEMENT CONTROL IN BUILDING ELEMENTS

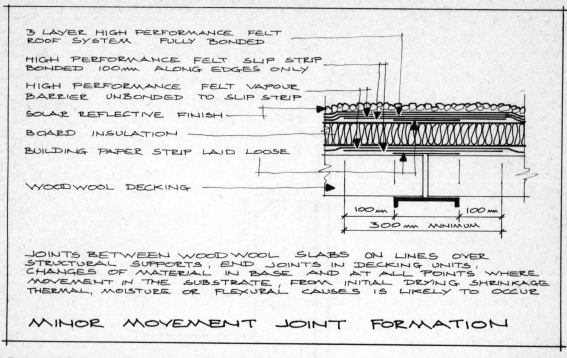

Diagram 5.7.13 Based on PSA *Flat Roofs* Technical Guide

Other serviceability limits to deflection

The various types of finish to floors and sheathing can impose other limitations, eg ISO 4356 gives a limit for asphalt or paved coverings of span/250 A similar limit for sand and cement floor screed seems sensible and for sensitive equipment or the floor a maximum slope of 1:750.

Stiffness in cantilever slabs taking cladding is particularly important. Deflections should not exceed 1:250 to 1:500 depending on cladding.

The International standards organisation *Deformation of buildings at the serviceability limit* ISO 4356, also gives a range of values for deflection limits based on the integrity of claddings supported at floor edges of 1:250 span to 1/500 span depending on cladding detail.

Visual limits

Visible deflections are not merely unaesthetic but can be seriously disturbing to the building users.

It is commonly accepted that deflections greater than *1:400 of* beam span and above span *1:180 m* cantilevers are visible and deflections should be kept below these figures.

Horizontal movement effect on other elements

Horizontal movement can seriously affect adjacent columns (see diagrams 5.8.2 and 5.8.3).

The effect of restraint particularly if asymmetrical can introduce a directional horizontal movement in floor slabs, (see diagram 5.8.2) provided this is anticipated and the movement is within the strain limitations of the construction. This will not cause failure unless a weakness, eg holes in floor, voids etc, are introduced into the critical areas, or where finishes are particularly sensitive to movement.

Horizontal movement can cause problems on external columns and cladding but also on internal columns (see diagram 5.8.3) particularly where movement in the roofs as shown in diagram 5.8.2 in the opposite directions increases the deformation of the affected members. The example in diagram 5.8.3 shows the increased effect if floor levels vary.

INTERNAL SUSPENDED FLOORS

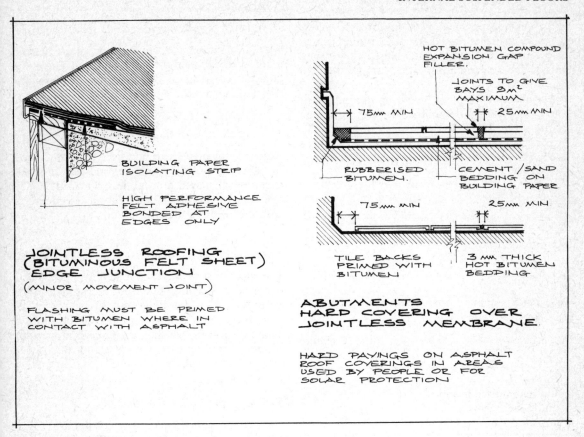

Diagram 5.7.14 Based on PSA *Flat Roofs* Technical Guide

Differential movements

Limits on angular deformation and localised induced stresses due to difficult support conditions described in 5.8.3 must be observed to prevent local cracking failures or partitions or other elements, eg cladding in contact with the floor.

CAUSES OF MOVEMENTS/ DEFORMATIONS IN FLOORS

Extrinsic causes: due to loading

Loading initial deflection:
The increased use of discontinuous pre-cast floor units and prestressed units has led to greater slenderness of members and greater deflections (see diagrams 5.8.4). This applies to floor types (1) and (2).

Initial deflection

Deflections due to dead partitions, etc, and live loading are the main cause of deflections. However some intrinsic causes can contribute. See below. Lack of stiffness is caused by large span depth ratios.

Deflections in reinforced concrete structures have increased because:

1. Precasting has led to loss of continuity
2. Prestressing itself can lead to higher deflections (see diagram 5.8.4A).

Note If prestressed members are prevented from moving, tension and increased compression can lead to failures (see diagram 5.8.4B).

Time related deflection: creep

Although caused by loading this is an intrinsic

effect. Creep occurs mainly in reinforced concrete structures and must be taken into consideration as an addition to the calculated elastic deflections, particularly if sections are cantilevered or exposed.

The effect of creep can be felt up to a period of one to two years after initial loading. Although initially caused by loading this is dealt with under intrinsic causes, since it also depends on the properties and type of concrete used.

Table 5.8.1

CP 116 1969 Clause 314 table 12
Span depth ratios:

Precast concrete

Maximum values of span/depth ratio of beams and slabs

	Ratio of span to overall depth
BEAMS	
Simply supported beams	20
Continuous beams	25
Cantilever beams	10
SLABS	
Slabs spanning in one direction, simply supported	30
Slabs spanning in one direction continuous	35
Slabs spanning in two directions simply supported	35
Slabs spanning in two directions continuous	40
Cantilever slabs	12

Note 1 The greater deflection of cantilevers is shown by these values.

Note 2 It should be noted that two way spanning slabs are subject to large deflections at the centre (see diagram 5.8.1). The general reinforced concrete crack width limitation applicable for interiors (ie *not* exposed to moist or corrosive atmospheres would also be applicable, ie CP 110 limit is 0.3 mm maximum crack width.

Differential support conditions and spares

Any local differences in floor or beam spaces, loading to differential deflections can affect the adjacent elements of construction, eg partitions (see diagram 5.8.5).

Effect of adjoining elements:

Adjoining elements can impose movements onto the slab and cause a significant increase in stress over the calculated normal design loading. They can also give differential restraint and cause induced stresses and deformations.

Temporary loads during construction

The effect of superimposed loads from props and upper floors can impose serious excess loads onto lower floor slabs, (see diagram 5.8.6).

As these imposed loads are usually put onto a member early in their life, extra creep movement will result, (see time of loading relationship under section 3 on concrete).

Extrinsic causes other than loading:

Since floors are not usually exposed to the *external* climate, interior temperature fluctuations are usually between 5°C to 25°C at the highest.

Differential temperatures

However, due to location in the buildings differentials between floors can arise (see diagram 5.8.5).

Intrinsic causes: shrinkage

Initial shrinkage of concrete:

In view of the long horizontal dimension of floors, and generally lack of horizontal restraint, shrinkage movements are likely to require careful consideration. The direction of movement will depend on the location of restraint as was shown in diagram 5.8.3. The amount of movement will depend on the quality of the concrete mix and all factors which were explained in section 3 on concrete. Other concrete early movements could

INTERNAL SUSPENDED FLOORS

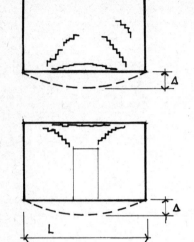

SAGGING: EXCESSIVE DEFLECTION
($\frac{\Delta}{L} > \frac{1}{300}$) OF SUPPORTING BEAM OR FLOOR

SCHILD RECOMMENDS NO CRACKING IF SPANS KEPT TO 4.5m – 7m MAX. FOR RIGIDITY IN FLOOR SPAN

SAGGING OF SUPPORT: EFFECT ON WALL AND OPENING

FOR FULLER TREATMENT SEE PARTITION SECTION 5.5

FLOORS: FAILURE LIMITS OF FLOOR CONTROLLED BY EFFECT ON ADJOINING ELEMENTS

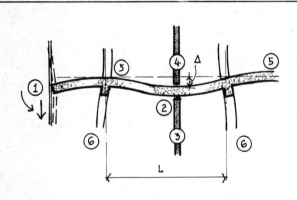

1. COMPRESSION IN CLADDING
2. TENSION IN CEILING PLASTER NOT CRITICAL IF CEILING IS SUSPENDED
3. COMPRESSION ON PARTITIONS BELOW
4. SAGGING IN PARTITIONS ABOVE
5. TENSION IN FLOOR FINISH
6. TENSION AND COMPRESSION STRESSES IN COLUMNS

DEFLECTION IN FLOOR SLABS: EFFECTS ON ADJOINING ELEMENTS

Diagram 5.8.1

DESIGN FOR MOVEMENT CONTROL IN BUILDING ELEMENTS

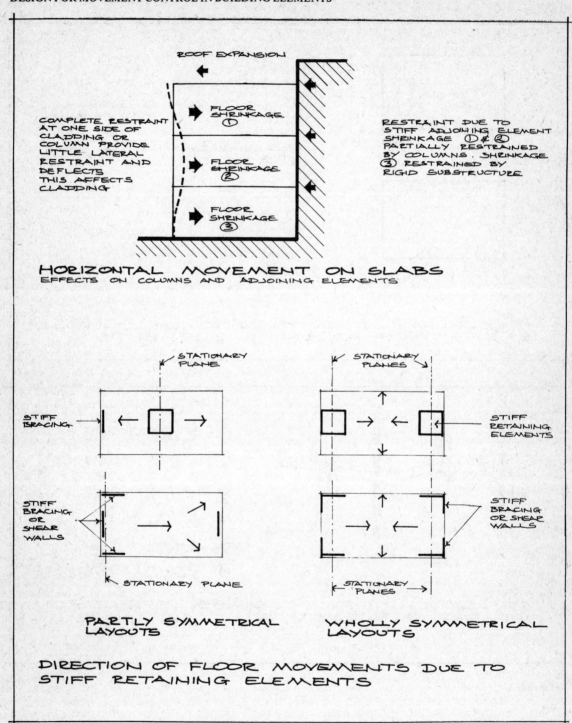

Diagram 5.8.2

INTERNAL SUSPENDED FLOORS

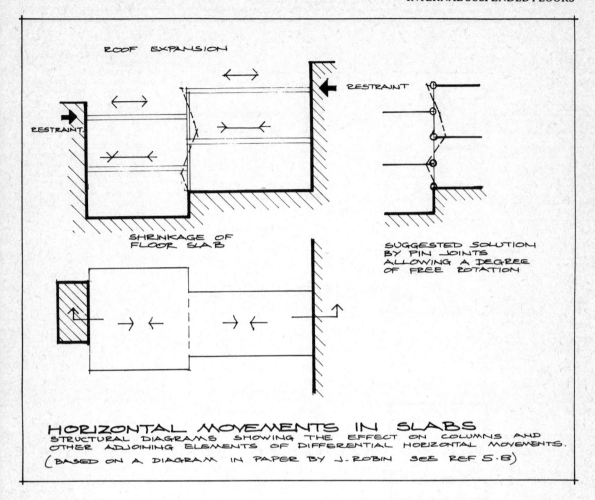

HORIZONTAL MOVEMENTS IN SLABS
STRUCTURAL DIAGRAMS SHOWING THE EFFECT ON COLUMNS AND OTHER ADJOINING ELEMENTS OF DIFFERENTIAL HORIZONTAL MOVEMENTS. (BASED ON A DIAGRAM IN PAPER BY J. ROBIN SEE REF 5.8)

Diagram 5.8.3

also take place, eg plastic settlement and initial shrinkage.

Internal floors should only be subject to initial drying shrinkage, since they are not subjected to alternate wetting and drying, except during construction, when they could also be subject to temperature changes greater than in the completed building. The initial drying shrinkage can last for a considerable time, months even years in some cases.

Time related movements in concrete floors

Creep movement of concrete slab as well as shrinkage movement will take place over a long period, as distinct from initial immediate deflation. This can be accommodated by the joints in a masonry partition whilst still plastic, whereas the later creep movement can cause cracking of the mature masonry.

Differential exposure conditions

This is particularly serious where slabs are exposed on one side externally and the other internally. However different temperature conditions can occur in various parts of a building. The top floors being more exposed to solar heat. Higher thermal insulation may not entirely overcome the heat build up in the structure.

179

DESIGN FOR MOVEMENT CONTROL IN BUILDING ELEMENTS

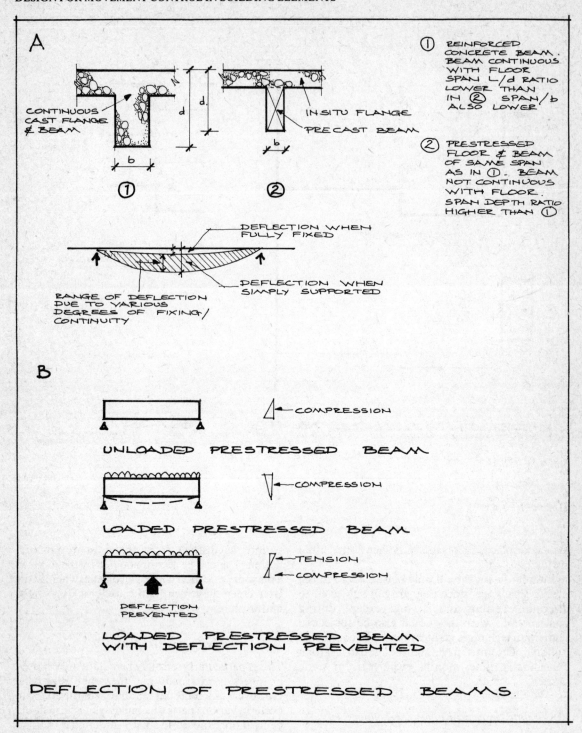

Diagram 5.8.4

INTERNAL SUSPENDED FLOORS

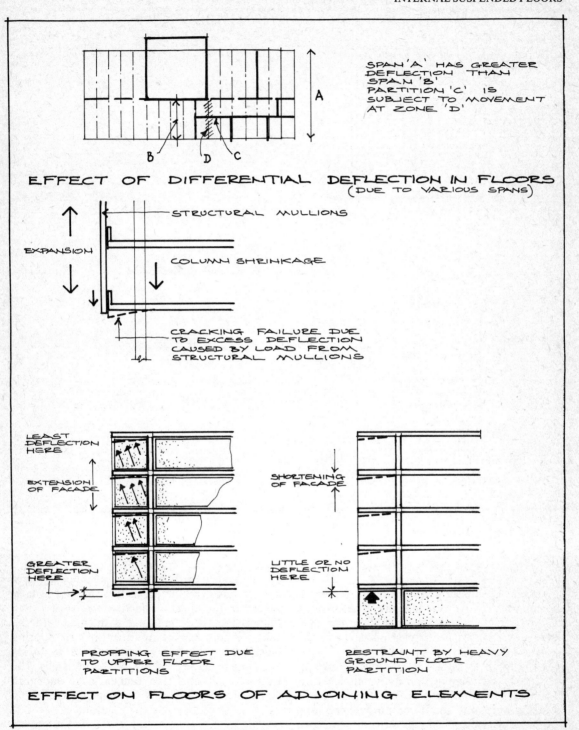

Diagram 5.8.5

DESIGN FOR MOVEMENT CONTROL IN BUILDING ELEMENTS

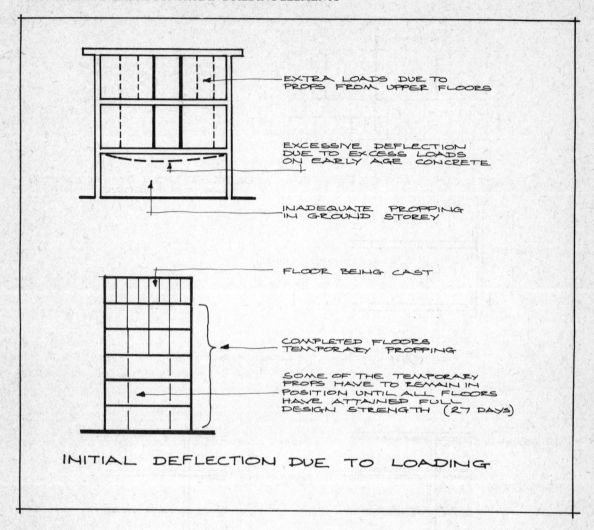

Diagram 5.8.6

Restraints

Column and beam supports and reinforcement in the slab will provide some restraint. The direction of this restraint can, as shown in diagram 5.8.3, cause movements to be directed unsymmetrically across the floor.

It is difficult to assess the amount of restraint provided by such supports precisely, but field tests have shown that the actual measured movements of a slab were only about one third of the total calculated unrestrained movement due to shrinkage and temperature changes.

This occurs in timber floor construction, but usually affects the boarding or strip flooring and not the carcassing timber as these are not usually subject to the moisture fluctuations. However in timber floors the boarding is integral with the supports and cases of movement in strip flooring have been published which have led to forces adequate to cause cracking in adjacent walls. The Codes of Practice for timber given moisture content levels given below to mitigate this effect.

Intrinsic movements in concrete

These have been explained in the section on

INTERNAL SUSPENDED FLOORS

concrete as a material. As slabs are the most vulnerable element using concrete due to the large surface exposed to drying and early temperature fluctuations, precautions for avoiding early and subsequent movement failures are given in this section as follows:

Floors generally

Whatever the material, the control of movements requires definite design precautions which can be summarised under the following headings and these are dealt with in more detail under each main type of floor construction:

Reduction of extrinsic movements

(a) Prevention of excessive deflections by adequate stiffness (span/depth ratios).
(b) Prevention of unequal deflections and induced stresses by unequal support conditions or loadings, without adequate stiffness, or separation by joints.
(c) Prevention of imposed deformations from adjoining elements, by provision of suitable separating joints.

Reduction of intrinsic movements

(a) Moisture content control in timber.
(b) Adequate curing and mix design to prevent excessive creep in concrete.

Movement joint provision

Where movements cannot be reduced to acceptable levels jointing will be required.

TIMBER FLOORS

To avoid excessive deformation in timber floors, the following main precautions are necessary:

Specification of timber

CP 112 Part 2 1971 quotes the mean and minimum modules of elasticity E for all grades of softwoods and hardwoods. This varies widely, e.g.

	Strength	$E\ (N/mm^2)$
Douglas Fir	6.6 – 18.6	6.600
European Spruce	3.4 – 11	3.800

Design for the correct grade actually used is not only vital for strength but for *stiffness*.

Structural design: slenderness limits

To prevent deformation due to slenderness of beams and joists CP 112 also gives maximum depth to breadth ratios for solid and laminated members, eg:

Degree of lateral support	Maximum depth to breadth ratio
Normal flooring	
No lateral support	2
Ends held in position and member held in line by purlins	4
Ends held in position and *compression* edge held in line by direct connection of sheathing deck or joists, together with adequate bridging or blocking spaced at intervals not exceeding $6 \times D$	6

Both slenderness and duration of loading are taken into account by the code by limiting factors or maximum stresses.

Workmanship

Cutting of notches should be limited to areas given in CP 110. End bearings should be constructed with care (see diagram 5.8.8).

Other precautions: settlement

Timber floors being discontinuous with surrounding structures and relatively flexible are not usually seriously affected by movements in adjacent structures. Only serious settlements in walls can lead to inequalities of level which can have disturbing effects without failure in the structure.

DESIGN FOR MOVEMENT CONTROL IN BUILDING ELEMENTS

CONCRETE FLOORS

Types (1) and (2) floors described earlier, being simply supported, are more liable to larger deflections than continuous type 3 floors.

Precautions require to be taken in design and in the workmanship during construction as follows:

Summary of design precautions:

1 Inequalities of support and 'propping' effect of partitions should be avoided. As illustrated earlier, these effects can cause slabs to deflect unequally and so cause failures in internal partitions which they support.

Reinforcement: design and workmanship

The structural design of the floor should incorporate sufficient areas of steel to limit cracks to 2 mm in width. This should be carefully located and prevented from moving during casting.

Mix design

To avoid excessive initial drying shrinkage, the precautions specified in the section on concrete should be taken.

Workmanship and early age movements

Plastic settlement shrinkage: protection from premature drying out and early vibration, particularly where unequal cross sections occur.

Failures due to Heat of hydration and early thermal cracking particularly in casting adjoining sections must be prevented.

Adequate compaction to avoid weakened sections due to loss of liquids.

Shuttering and effect on deflection and creep

Shuttering should be strong enough to prevent deflections due to the weight of wet concrete (dry weight × 2).

Striking too early and inadequate propping, leads to early age loading with extra creep deflections.

FLOOR MOVEMENT JOINTS: DESIGN PROCEDURE

Provision of joints

1 Any such joints must be carried through the entire structure, including floors, so that each side of the joint slabs will need independent support details as shown in diagrams 5.8.7A and 5.8.7B.

Minor movement joints

2 Any horizontal movement control joints for control of shrinkage or thermal movement only require free horizontal slab movement of the slab both at the joint and at Wall bearings, or other restraints.

Functional requirements of floor joints

Although initially designed to accommodate movements, the selection and detailing of floor joints is carried out with due regard to the following criteria and additional requirements:

1 Requirements of finish and joints in finishes:
 Waterproof: cleaning materials
 Edging of finish
 Acceptable appearance
 Traffic: pedestrian traffic
 vehicular traffic
2 Structural requirements
 Support of slab edges at joint
3 Continuity with wall joints
4 Other floor functions:
 Fire/acoustic resistances
 Fire resistance tests on jointing is currently proceeding
5 Direction/size of movement
 Horizontal or vertical
 40 mm or wider joints require special components as in bridge construction. Link slabs may require rotational movement
6 Appearance on ceiling below or if not critical, eg false ceiling below
 Joints in plaster finish.

Location of floor joints spacing

Full floor structural movement joints are required in the following locations (see diagram 5.8.9).

INTERNAL SUSPENDED FLOORS

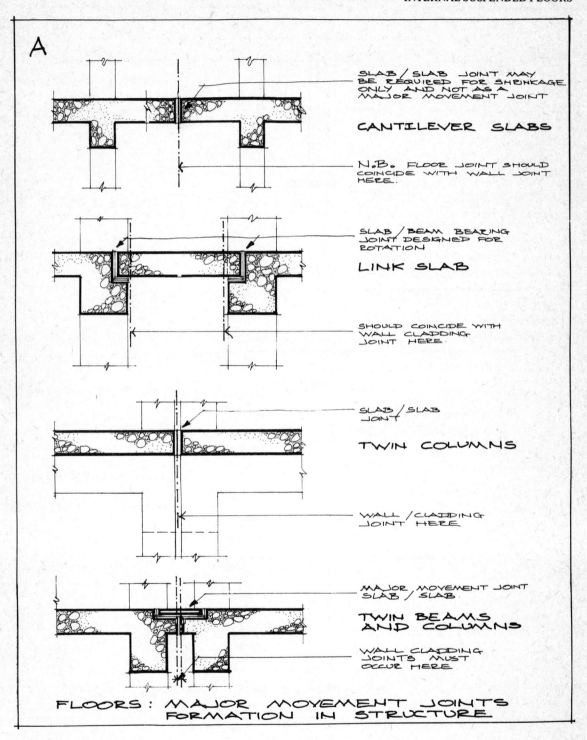

Diagram 5.8.7A

DESIGN FOR MOVEMENT CONTROL IN BUILDING ELEMENTS

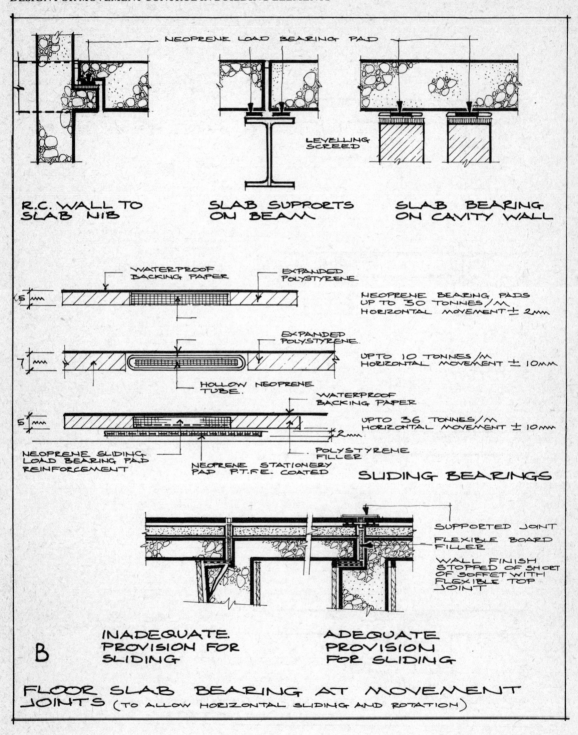

Diagram 5.8.7B

INTERNAL SUSPENDED FLOORS

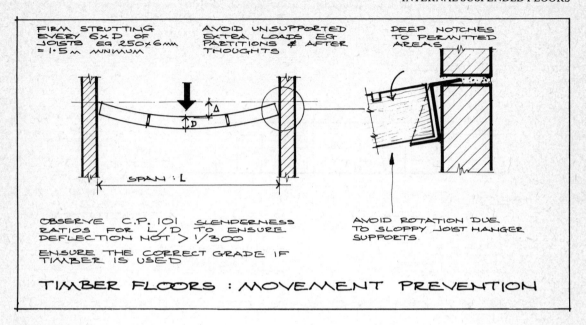

Diagram 5.8.8

1 To coincide with joints in structural frameworks. See section on frameworks, eg link slabs.
2 In buildings over 30 m long.
3 Where induced stress concentrations are likely to occur (see also diagram 5.8.9)

Location in structure

Floor joint types: (see diagrams 5.8.9 and 5.8.10).
Floor slab – floor slab
Floor slab – bearing masonry
Floor slab – beam bearing
Slab and beam – slab and beam
Floor beam – supporting beam bearing.

Movement type accommodated

Minor joints for accommodation of horizontal movement only.
Major movement joints (formerly termed expansion joints) to coincide with structural separations for settlement, vertical horizontal and rotational movements.

Traffic type

Light pedestrian
Heavy pedestrian
Vehicular and heavy duty.
(see 5.8.5 (b))

FLOOR JOINTS SEALANTS AND FILLERS: MAIN TYPES

Types of seal and joint fillers (see diagram 5.8.10)

Finishes	Jointing in structure (slab)	Soffit	Use
Flexible sealants	Flexible filler	Cover plate Capping strip	For pedestrian wear water resistant
Sprung cover strip	Void	Timber cover plate	Pedestrian wear not water resisting
Moulded gasket supported by extruded metal section			Vehicular traffic water resistant

DESIGN FOR MOVEMENT CONTROL IN BUILDING ELEMENTS

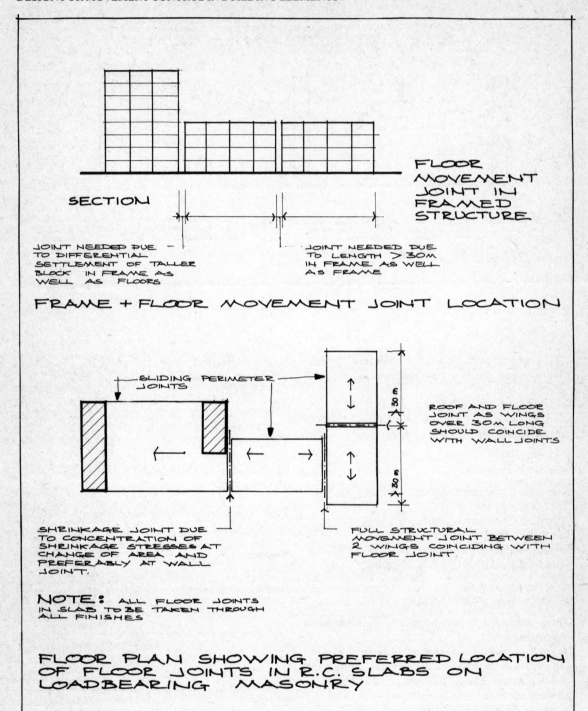

Diagram 5.8.9

INTERNAL SUSPENDED FLOORS

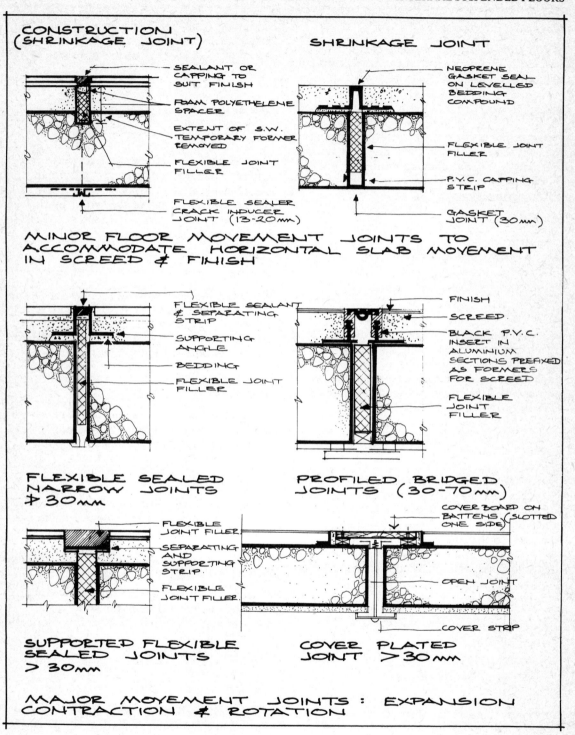

Diagram 5.8.10

DESIGN FOR MOVEMENT CONTROL IN BUILDING ELEMENTS

Floor jointing components

The flexible filling and sealing compounds are selected in accordance with the principles given in the section on sealants.

Finishes and supports

However floors present special problems due to the need to resist wear at the joint in the finishes and need to present a good appearance at soffits.

Special components have been developed and are shown in diagram 5.8.10.

5.9 INTERNAL FINISHES

FLOOR FINISHES

Sub-floor system

Floor finishes are an integral part of solid floor design and the substrate and base design will be affected by the requirements of the finish, for degree of level, water resistance and tolerance of movements. All major movement joints in the base must be carried through to the finish.

SCOPE

This section will deal mainly with hard brittle floor finishes particularly if unjointed such as granolithic and terrazzo as these have little movement tolerance, either to linear movement or to bending, ie deflection. This could be due to loading or warping or curling of screed or slab sub bases. Jointed finishes, if cement bedded are none the less sensitive to movements.

DESIGN CRITERIA: MOVEMENT FAILURE LIMITS

Variable requirements of finishes

Various floor finishes have greatly different failure limits depending on their form, method of laying, bedding, and constituent material properties.

Hard, cementations and jointless *in situ* cast finishes (see CP 204 *in situ floor finishes*) have to be laid in bays) and any cracks greater than 0.2 mm (ie hair cracks in the base) would be transmitted to and could be greater than the resistance of the screed.

Flooring assembly types

Solid bedded floor finishes and finishes supported on a suspended floor or suspended spaced supports have very different resistance to movements and their requirements vary accordingly. In solid *bonded* screeds properly laid, there is a resistance to shrinkage of the screed by the sub base. If unbonded, there is lesser resistance to shrinkage in the screed.

Cracking

This is due to linear movements or other dimensional deformation, such as differential vertical movements on excessive deflection greater than $>1/300$ span on the floor. Again failure limits vary greatly between hard and soft finishes.

Detachment

Minor movements of the subbase can lead to detachment or loss of bond/adhesion of tiles or jointed units due to loss of bond.

Curling/warping of screeds

Curling and warping of screeds leads to breakdown of the screed itself and is unacceptable in the finishes.

Moisture movement in timber

When flexible flooring is laid over jointed timber boarding, the normal slight moisture movements and warping of individual boarding normally acceptable will deform the finish and this movement becomes unacceptable.

Timber deflection

The slight deflections in timber floors even if less than $1/300$ span would crack rigid tile or cementitions finishes, unless special precautions are taken.

INTERNAL FINISHES

Moisture movement failures

The pressure of moisture in the subfloor and fluctuations in the moisture content of timber floorings can cause unacceptable movements in the timber flooring, so that in the case of timber floors, whether timber strip over a void, or solid bedded, moisture levels should be carefully controlled.

CAUSES OF MOVEMENT AND EFFECTS

These can be attributed to movements in the finish itself or induced movements transmitted from the subfloor or base.

Table 5.9.1 is a list of examples from each type of flooring, rather than a comprehensive list of hard cement based or mortar bedded floorings:

Table 5.9.1

Flooring	Material	Cause	Effect
Hard: jointless: finishes	Cement/ sand screeds	Deflections in base Shrinkage in base Initial drying shrinkage	Cracking Detachment break-up Crazing Curling Warping as per screeds
	Granolithic or concrete *in situ*	As per screeds	
	Terrazzo	As per screeds but also aggregate shrinkage	As per screeds
Hard: jointed finishes	Ceramic/ clay tiling	Shrinkage in bedding differential movement to screed sub-base by thermal movements. Moisture expansion of tiles.	Detachment 'arching' 'lifting' Detachment cracking joint failure

FAILURES IN FLEXIBLE FLOOR FINISHES

Flexible floorings are however not immune from failure due to movements and these are summarised below:

Timber flooring

Both s.w. boarded or narrow strip timber floors are seriously affected by moisture content fluctuations as shown in the section on materials. Excessive loads on adjoining walls have caused structural damage as well as deformation in the flooring.

Wood block or solid bedded finishes

have a very low tolerance of moisture changes before lifting or arching takes place.

Thermoplastic sheet/tile flooring

can be detached by expansion of trapped moisture vapour causing rippling, bubbling and consequent cracking at high points.

MOVEMENT CONTROL METHODS IN FLOORS

Alternative Strategies

As in other elements there are two types of design precaution to be taken:

1 Prevention or reduction of movements by design and workmanship
2 Control of effect of movement by *separation* and jointing and by *restraint* or reinforcement.

For ease of reference the following summary of precautions to be taken gives both types of approach for each material.

Cement based in situ *floor finishes*

Cement/sand screeds

Movement reduction/prevention

1 Provide a firm base (1:2:4 concrete mix), well compacted on a firm base.
2 Preparation of base surface, clean free from

191

dust and debris and roughened surface to achieve best bond with
3. lay screed in 15 m² bays not $> 1\frac{1}{2} \times 1$ length/width
4. Low water cement ratio of mix. Use a *coarse*, ie sharp sand.
5. Adequate curing essential.
6. Adequate thickness to develop strength against drying shrinkage, (see diagram 5.9.1)

Accommodation of movements

Any joints in the sub base/floor must be taken through the screed also as this will not accommodate movements requiring joints in the subfloor base.

Joints at perimeter may be advisable in heated screeds only. In normal screeds, a shrinkage crack at the perimeter is acceptable, unlike in grano or terrazzo floors where an *in situ* skirting is formed.

The control of movement in cement/sand screeds also depends on the movement characteristics of the finish to be applied over the screed itself, so that cracking is controlled within the limits of the finishes shown in table 5.9.2.

Terrazzo

(a) *Movement reduction/prevention/restraint*

1. *Mix design not* >1 cement to 2 aggregate by volume. Correct grading to avoid wet harsh mix, (ie not < 3 mm) keep water/cement ratio low.
2. Lay monolithically with base (first 12 mm), ie three hours after casting of the dense base screed to achieve shrinkage resistance by restraint.
3. Curing: during laying protect from high temperature, sun, droughts/wind. Leave covered 24–48 hours after laying.

(b) *Movement control*
Use ebonite, plastic or metal dividing strips. Set into screed before placing terrazzo with area not > 1 m² and 3:1 proportion.
In addition a strip round perimeter and all obstructions/holes is necessary. Panel sizes do not vary with the method of forming the topping but thicknesses do, so CP 204 should be consulted in detail.

Table 5.9.2

Type of floor covering	Type of screed construction	Thickness mm	Maximum bay sizes (between joints)
Thin sheet or PVC tile, or carpet (thin finishes show any curling in screed first class bonding required)	Bonded	20–40	Maximum width 4.5 maximum any length
Strong tiles, eg concrete or ceramic	Bonded	20–40	Width max 4.5
	Partially bonded	over 40	Width max 4.5
Thin covering	Heated subfloor unbonded	60 mm minimum	Width 4.5 max/15 m²
Strong covering	on insulation		panels near square

[1] Table based on C and CA *Concrete ground floors* by R Colin Deacon. This should be consulted for further details.

Ceramic/clay tiles

(a) *Movement reduction/prevention*
differential movement between tiles and the screed which has a greater drying shrinkage cannot be prevented. Thermal movements in the concrete base will obviously be less than the finish of tiles.
To prevent these movements being passed from the tiles to the base, two methods are used.

1. Bedding on a separating layer consisting of polythene sheet layer with lapped joints cement/sand bedding of at least 13 mm but preferably 19 mm.
2. Thick bed say 40 mm thick, ie virtually an independant screed.
3. Flexible adhesives to tolerate the differential movements between tiling and base.

(b) *Movement control*
The above measures are not adequate except

INTERNAL FINISHES

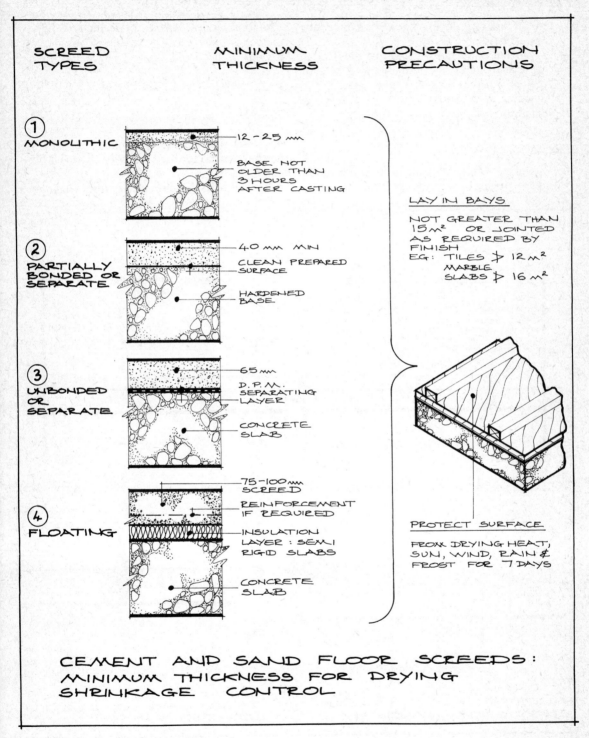

Diagram 5.9.1

DESIGN FOR MOVEMENT CONTROL IN BUILDING ELEMENTS

over a relatively small area, and certainly not against hard abutments.

It is therefore necessary to provide flexible movement joints in floor tiling at 4.5 m centres, at the perimeter and against the junction with abutments – as shown in diagram 5.9.2.

Wood block and strip flooring

1 *Prevention* Moisture content control prior to and after laying in wood block floors, provision of joint around perimeter (see diagram 5.9.2).

2 *Control*

INTERNAL WALL FINISHES

SCOPE

As with floor finishes, applied Internal wall finishes depend for their integrity on the wall base and background.

All wall joints should be carried through the finish.

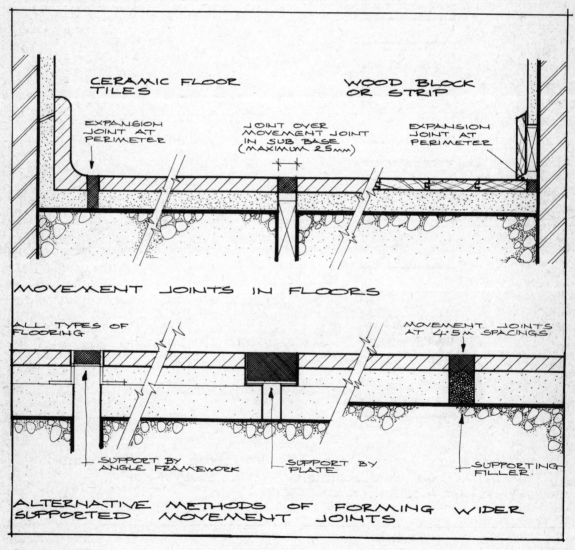

Diagram 5.9.2

INTERNAL FINISHES

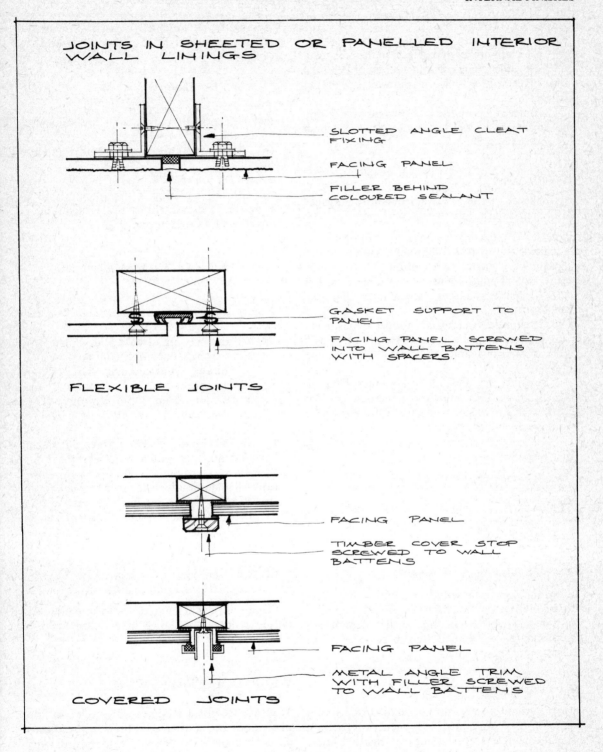

Diagram 5.9.3

DESIGN FOR MOVEMENT CONTROL IN BUILDING ELEMENTS

DESIGN CRITERIA

Since even hair cracks and defects in the finish will be visible, unless they occur in a recess or behind a cover plate, failure limits for thin wall finishes are critical.

The objective therefore is to prevent or hide any movement cracks which cannot be covered by a decorative paint film.

Finishes: types

From the view point of movement control, finishes can be regarded as falling into two types.

Type 1   Dry fixed linings over framing/joists or jointed and panelled finishes. Although still requiring a degree of attention to movements this type of panelling usually requires attention to the panel material itself, rather than the background.

Type 2   Jointless wet directly applied to base such as plasters or renderings. These are obviously the most critical and will be dealt with in some detail.

So called 'dry' plasterboard linings still have a measure of direct adhesion to the base are still vulnerable to induced movements from the base and more liable to failure.

CAUSES OF MOVEMENT

All movements causing cracking in the wall base as described in section 5.5 will cause cracking in the finishes particularly if type 2 as above.

Extrinsic differential movements

In addition, differential backings (eg RC frame/infill panel walls) will cause differential movements which the thin finishes cannot accommodate, so that design precautions are necessary at these points.

Intrinsic movements

Most gypsum plasters do not suffer from drying shrinkage and subsequent movements after setting are small, but too rapid drying can lead to delayed expansions.

Cement (sand rendering) has to be carefully specified and carried out to avoid drying shrinkage problems.

Jointed: dry fixed finishes

1 *Prevention/reduction of effects*
   (a)   Selection of panels with low movement characteristics.
   (b)   Moisture content control before and after installation.
2 *Movement control*
   (a)   Flexible jointing at fixing points or allowance for movement by sliding junction and cover beads (see diagram 5.9.2).

MOVEMENT CONTROL METHODS: WET PLASTER FINISHES

1 *Reduction of effects/preventions on solid masonry base*
   (a)   Correct wetting of base and allowance of time for base to complete initial drying out shrinkage particularly lightweight blockwork.
   (b)   Protection from rapid drying out on *plasterboard fixed to timber framing/supports*.
   (c)   BC 5492–1977 gives guidance on spacing and moisture content requirements.
2 *Movement accommodation by joints*
   (a)   All major joints to be taken through finish.
   (b)   Provide movement joints at intersections between surfaces where relative movement may occur.
   (c)   Provide movement joints for all intersections of backing material.
   (d)   Cover all joints with joinery, or provide a stop, all as in traditional construction.

Details of typical methods for forming joints in plaster finishes are shown in diagrams 5.9.5 to 5.9.8.

WET BEDDED WALL TILING

Movement control methods are similar to floor tiling described earlier. Junctions with adjacent elements require movement joints (see diagram 5.9.7).

INTERNAL FINISHES

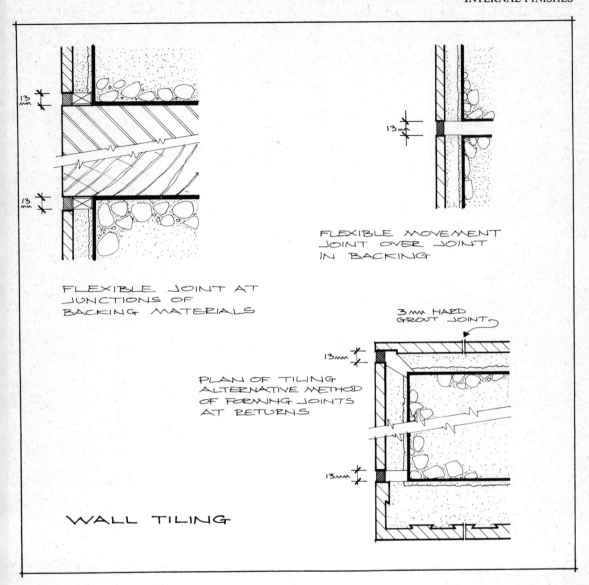

Diagram 5.9.4

FRAMING FOR DRY FIXED LININGS/PLASTERBOARD

Metal lathing should be as taut as possible; this is best achieved by nailing the sheet in the centre and then working towards each end. Ends of wire for fastening should be bent inwards and not towards the plaster coat.

After erection all cut edges and damaged metal lathing stapes or nail-heads should be given a protective coat of bitumen paint (see CP 231).

Timber supports for boards should be accurately spaced and properly aligned and levelled. In order to minimize the risk of cracking at the board joints, seasoned timber with a moisture content not exceeding that given in table 5.9.4 should be used. It is important that the dimensions and construction of timber supports should

197

DESIGN FOR MOVEMENT CONTROL IN BUILDING ELEMENTS

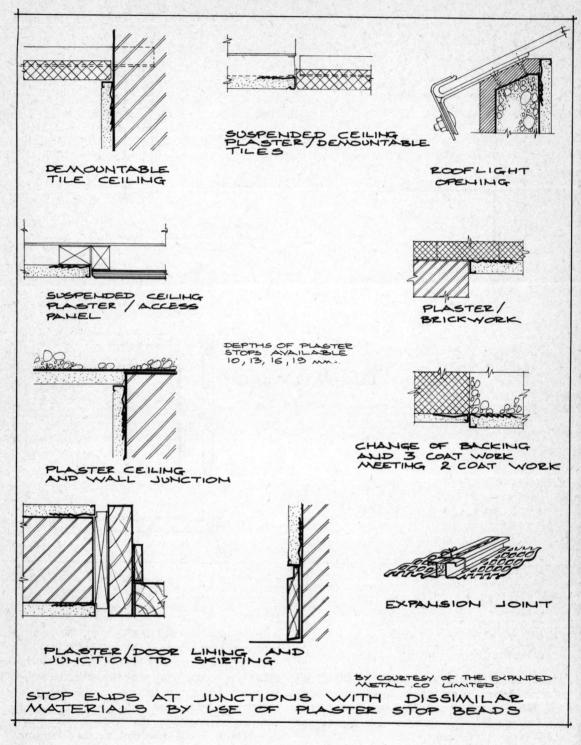

Diagram 5.9.5

INTERNAL FINISHES

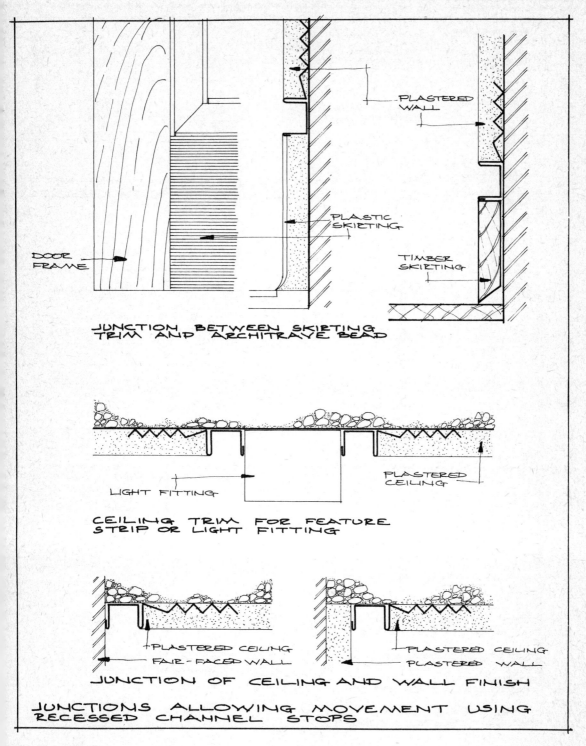

Diagram 5.9.6

DESIGN FOR MOVEMENT CONTROL IN BUILDING ELEMENTS

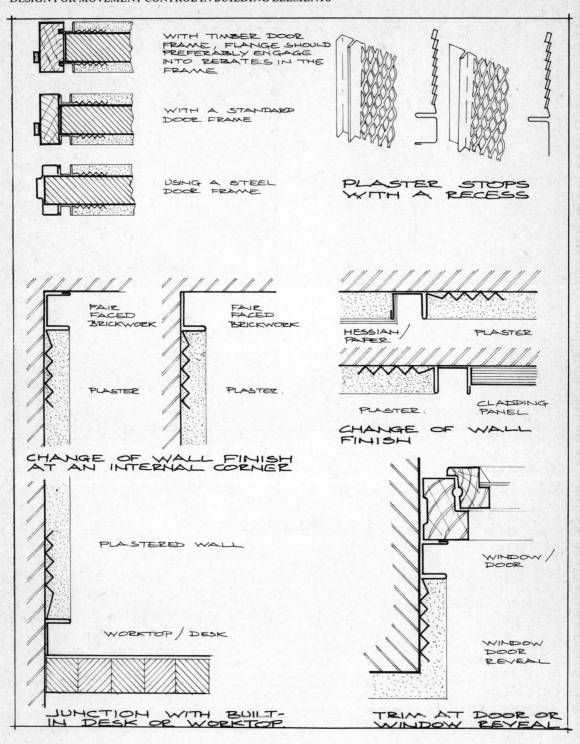

Diagram 5.9.7

SUBSIDIARY COMPONENTS

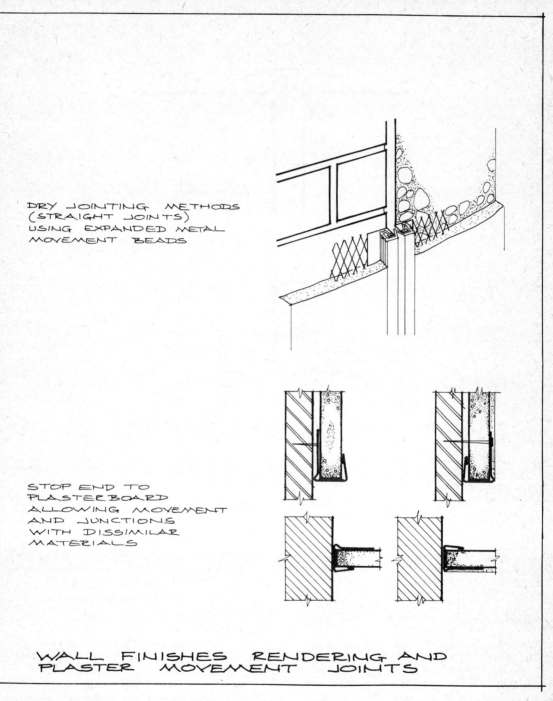

DRY JOINTING METHODS (STRAIGHT JOINTS) USING EXPANDED METAL MOVEMENT BEADS

STOP END TO PLASTERBOARD ALLOWING MOVEMENT AND JUNCTIONS WITH DISSIMILAR MATERIALS

WALL FINISHES RENDERING AND PLASTER MOVEMENT JOINTS

Diagram 5.9.8

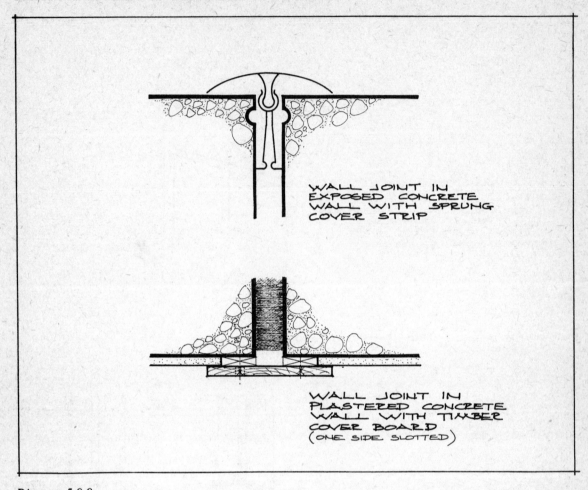

Diagram 5.9.9

Table 5.9.3

Supporting members	Board thickness	Recommended centres
Vertical: studs or furring	9.5 mm lath	450 mm
	9.5 mm baseboard	450 mm
	12.7 mm lath	600 mm
	19 mm plank	800 mm
Horizontal: joists or furring	9.5 mm lath	400 mm
	9.5 mm baseboard	400 mm
	12.7 mm lath	450 mm
	19 mm plank	750 mm

permit the nailing of two ends of boards to a support not nearer than 13 mm from the edges and allow the nails to be driven in straight, ie not skewed to a secure bed in the timber. The dimensions of the supports should also allow for a gap, not exceeding 5 mm, to be left at the end joints of gypsum lath baseboard and plank. Recommendations for spacing of supports are given in table 5.9.4

Where boards have to be fixed at centres greater than those recommended in table 5.9.4 the maximum spacings for various thicknesses of plasterboard should be as shown below.

Where boards are fixed at maximum centres it should be recognized that there is a greater risk of sagging and that there is no margin of safety to allow for adverse conditions.

(a) Maximum spacings for timber supports for

9.5 mm gypsum lath and baseboard should be 450 mm centres.
(b) Maximum spacings for timber supports for 12.7 mm gypsum lath should be 600 mm centres.
(c) Maximum spacings for timber supports for 19 mm gypsum plank should be 800 mm centres.

Moisture content of timber for various positions (extract from CP 112: part 2: 1971)

Table 5.9.4

Position of timber in building	Average moisture content attained in use in a dried out building (per cent of dry weight)	Moisture content which should not be exceeded at time of erection (per cent of dry weight)
Framing and sheathing of timber buildings (not prefabricated)	16	22
Timber for prefabricated buildings	16	17 for precision work otherwise 22
Rafters and roof boarding, tiling battens, etc	15	22
Ground floor joists	18	22
Upper floor joists	15	22

5.10 SUBSIDIARY COMPONENTS

SCOPE

Movement control in components, particularly in joinery are traditional. This handbook in concentrating on new problems refers the reader to the many excellent textbooks on joinery and *Mitchell's Materials* by Alan Everett for data on moisture movement control in timber. (See diagrams 5.10.1 and 5.10.2.)

Metal and plastic components, particularly in windows, are subject to very high temperature fluctuations and such systems should allow adequate tolerances and flexible or sliding joints both in the assembly and at junctions with the surrounding constructions. Both vertical mullions and long sills are particularly vulnerable.

DESIGN CRITERIA

GLAZING SYSTEMS

The greatest progress and changes have taken place in recent years in systems of glazing.

There has been a tendency to use larger panes and double glazing may increase movement problems. Fortunately more flexible systems of retention of the glass have been developed by the use of:

(a) flexible non-setting glazing compounds

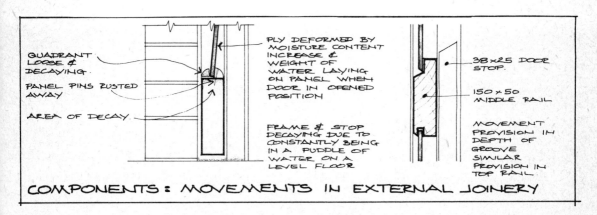

Diagram 5.10.1

DESIGN FOR MOVEMENT CONTROL IN BUILDING ELEMENTS

(b) flexible gaskets
(c) a combination of both.

The design of metal sections has also advanced considerably. Some typical examples are shown in diagrams 5.10.3 and 5.10.4 and subsequent illustrations.

However the following extract table from Messrs Tremco's handbook on glazing system illustrates the critical nature of the edge cover (see 5.10.8) for various sizes of glazing panel.

Minimum edge cover to glazing.

Edge cover

Glass area in M^2	Wind loading 1.000 N/M^2	1.500 N/M^2	Over 1.500 N/M^2
Up to 1.2	6	9	12
Up to 2.0	9	9	12
Up to 3.0	9	12	15
Up to 4.6	9	12	15

EXTERNAL JOINERY: DRAUGHT EXCLUSION

BSDD4:1971 for windows and doors requires specific performance standards under the following headings all related to specific wind pressures, based on degrees of exposure

1 Deformation (Deflection) due to wind loads
2 Air infiltration
3 Water penetration

As timber windows and doors are also particularly vulnerable to deformations due to moisture movement and are liable to failure under (2) and (3), various flexible joint sealing techniques have been developed to overcome the problem and are illustrated in diagram 5.10.2.

MECHANICAL SERVICES

Pipes

This is an area which may affect the building designer, since expansion joints in pipes can be space consuming, (see diagram 5.10.5) roller support to a heating pipe to allow freedom of movement.

Floor heating

Heated pipes buried in the construction will affect all adjoining finishes.

For example a heated screed covered in marble tiles has to be restricted to a maximum area between movement joints of 16 m^2.

Underground drainage

Movements in the ground so frequently the cause of drainage failures in the past, have largely been avoided by the use of flexible jointing methods and flexible bedding in fine pea shingle beds, (see diagram 5.10.6) and the use of flexible drain and sewer pipes.

SUBSIDIARY COMPONENTS

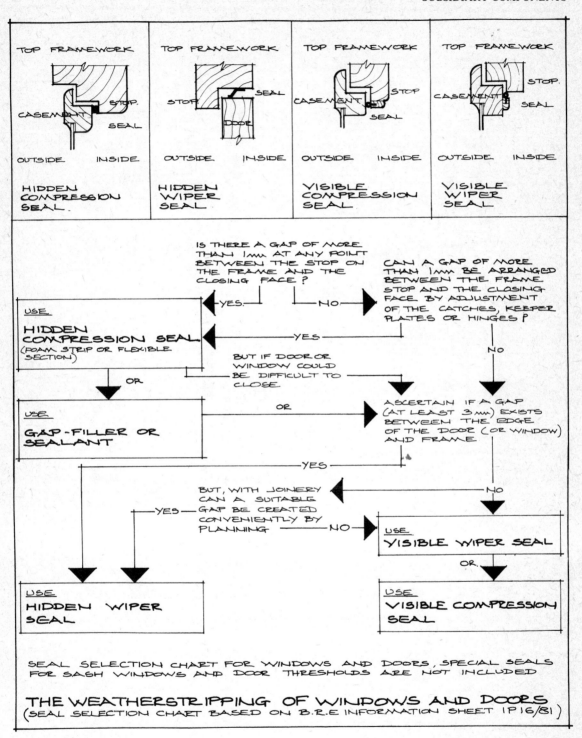

Diagram 5.10.2

DESIGN FOR MOVEMENT CONTROL IN BUILDING ELEMENTS

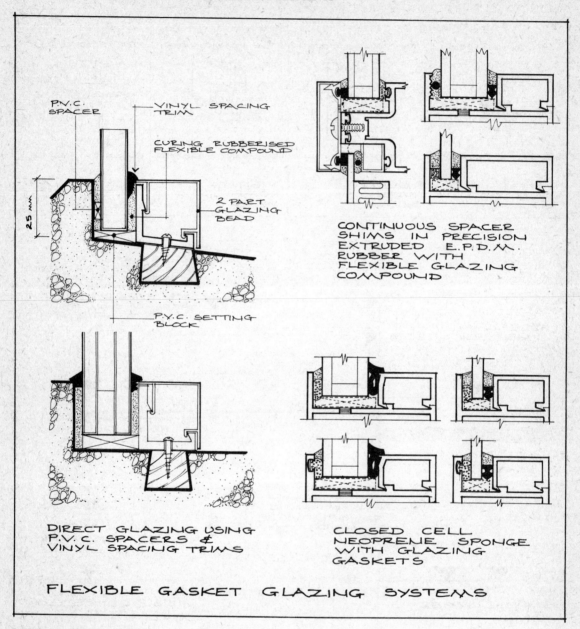

Diagram 5.10.3

SUBSIDIARY COMPONENTS

GLAZING: MOVEMENTS TO BE ACCOMMODATED

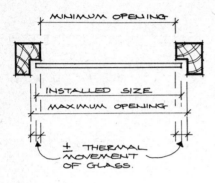

± THERMAL MOVEMENT OF GLASS.

BOWING/DEFLECTION DUE TO WIND GUST LOADS

NOTE 8
IF THE GLAZING SYSTEM CAN ACCOMMODATE BOWING SOME THERMAL EXPANSION CAN BE ACCOMMODATED AS WELL AS LINEAR EXPANSION

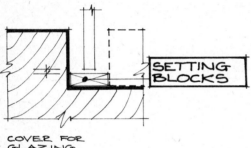

RESILIENT NON ABSORBING TO MAINTAIN AN EVEN CLEARANCE ALL ROUND FRAME AS IN B.S.C.P. 152 IN NYLON OR UNPLASTICISED P.V.C.

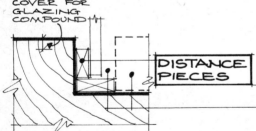

RESILIENT NON ABSORBING. SIZE 25mm × 3mm × WIDTH LEAVING COVER FOR COMPOUND SET 350mm MAX APART AND OPPOSITE ON EITHER SIDE OF GLASS.

SEALERS, GASKETS OR GLAZING COMPOUNDS 'FRONTING' WITHOUT BEADS OF FILLING WITH BEADS

GLAZING: ESSENTIAL COMPONENTS IN LARGER PANES OR DOUBLE GLAZING SYSTEMS TO ENSURE ADEQUATE EDGE CLEARANCES

Diagram 5.10.4

DESIGN FOR MOVEMENT CONTROL IN BUILDING ELEMENTS

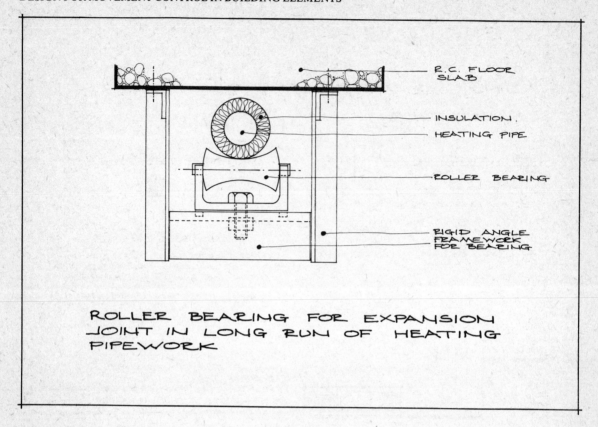

Diagram 5.10.5

SUBSIDIARY COMPONENTS

PIPE JOINTS

Typical joints for flexible drain and sewer pipes are shown opposite. Solvent-welded joints and pitch fibre tapered joints do not provide for telescopic movement.

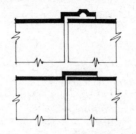

U.P.V.C. PIPES

- RUBBER RING COMPRESSION OR SEAL JOINT USING LUBRICANT
- SOLVENT WELDING JOINT

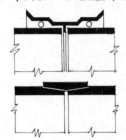

PITCH FIBRE PIPES

- POLYPROPYLENE SNAP RING JOINT (AVAILABLE FOR 100 & 150mm BORE)
- TAPER JOINT (AVAILABLE FOR 200 & 250mm BORE ONLY)

KEY TO TABLE BELOW

* U.P.V.C. PIPES TO B.S. 3505 & B.S. 3506 ARE ALSO SUITABLE BUT FITTINGS MAY NOT BE AVAILABLE, AND THE PIPES ARE NOT INTERCHANGEABLE WITH THOSE IN B.S. 4660 & B.S. 5481

S. NORMALLY SUITABLE

SS. SPECIALLY SUITABLE

X. EXPERT ADVICE SHOULD BE SOUGHT FROM SPECIALIST OR MANUFACTURER

TYPE OF PIPE	B.S. REFERENCE	NORMAL DOMESTIC SEWAGE	SOIL ENVIRONMENT CONTAINING SULPHATES & ACIDS	CHEMICAL RESISTANCE TO TRADE EFFLUENTS CONTAINING ACIDS AND ALKALIS.			
				ORGANIC SOLVENTS	VEGETABLE AND MINERAL OILS/FATS	AT NORMAL TEMPERATURE	AT SUSTAINED HIGH TEMPERATURE
*U.P.V.C.	4660 (FOR PIPES OF 110 & 160mm NOMINAL SIZE)	S	S.S	X	S	SS	X
	5481 (FOR PIPES OF 200mm NOMINAL SIZE AND ABOVE)	S	S.S	X	S	SS	X
PITCH FIBRE	2760 (STANDARD LIMIT OF SIZE IS 200mm NOMINAL BORE, BUT 225mm IS AVAILABLE)	S	S.S	X	X	S	X

Diagram 5.10.6

General references to all sections

**Section 2, Sources and effects of movement, and
Section 4, Design for movement in buildings**

Alexander, S J and Lawson, R M, *Design for Movement in Buildings, CIRIA Technical Note 107*, CIRIA, 1981

Bickerdike, A Rich and Partners, *Buildings*, first and second series, George Godwin, 1972

BS 5606: 1978, *Code of Practice for accuracy in building*, BSI, 1978

BS 6093: 1981, *Design of joints and jointing in building construction*, BSI, 1981

BS 79/10968, Third draft *Standard guide to the selection of constructional sealants*, BSI, 1979

BRE, *Building Materials: essential information from the Building Research Establishment*, MTP Construction, 1973

BRE, *Building defects and maintenance*, The Construction Press, 1977

BRE CP 2/73, *Accuracy and its structural implications for loadbearing brick constructions*, Milner, R M and Thorogood, R P, BRE 1973

BRE Digests 227–229: 1979 *Estimation of thermal and moisture movements and stresses*, Parts 1–3, HMSO, 1979

BRE Digest 199: 1977, *Getting a good fit*, HMSO

BRE Digest 75: 1966, *Cracking in buildings*, HMSO, 1968

BRE Digest 251: 1981, *Failure patterns and implications*, HMSO, 1981

BRE Digest 223: 1980, *Wall cladding: designing to minimise defects due to inaccuracies and movements*, HMSO, 1979

BRE Digest 137: 1977, *Principles of joint design*, HMSO, 1972

BRE CP, *Tolerances and fits for standard building components*, Ronshor, R B and Eldridge, L, Distribution Unit, BRE, 1974

BRE CP 99/75, *Accuracy of in-situ concrete*, Thorogood, R P, Distribution Unit BRE, 1975

BRE CP 29/70, *Weatherproofing of joints: a systematic approach to design*, Harrison, H W and Bonshor, R B, Distribution Unit, BRE, 1970

BRE CP 5/71, *The relationship between component size and joint dimension*, Bonshor, R B and Harrison, H W, Distribution Unit BRE, 1971

The Concrete Society, *Design for movement in buildings*, Proceedings of a symposium by the Concrete Society, The Concrete Society, 1969

Ellis, M, Hutchinson, B D and Barton, J, *Maintenance and repair of buildings*, Newnes-Butterworths, 1975

Hart, F, Henn, W and Soutag, H, *Multi-storey buildings in Steel*, Granada Publishing, 1978

The International Symposium for the Construction Industry, *Joint Movement, Design and Materials*, Conference Papers of the Sealants Manufacture Conference, May 1970

Lenczner, D, *Movement in Buildings*, Pergamon Press, 1981

Martin, Bruce, *Joints in Buildings*, George Godwin, 1977

Miles, D, *A manual on Building Maintenance*, vols 1 and 2, Intermediate technology Publications, 1979

NBA, *Common Building Defects: diagnosis and remedy*, The Construction Press, 1979

Neville, A M, Houghton-Evans, W and Clarke, C V, *Deflection Control by Span/Depth Ratio*, C & CA magazine of Concrete research, 1977

Schild, E, Oswald, R, Rogier, D, Schweikert, H, Schnapauff, V, *Structural failure in Residential Buildings*, vols 1 to 4, Granada Publishing, 1979

The Sealants Manufacturers Conference, *Manual of good practice in sealant application*, The Sealants Manufacturers Conference, 1979

Trill, J and Bower, J T *Problems in Building Construction. A Scientific Method Approach*, Book 1, Architectural Press, 1972

Section 3, Movement characteristics of materials

BS CP 110 Part 1: 1972, *The structural use of concrete: design, materials and workmanship*, BSI, 1972

BS CP 116: 1969, *The structural use of precast concrete*, BSI, 1969

BS CP 297: 1972, *Precast concrete cladding (non-bearing)*, BSI, 1972

BS CP 121 Part 1: 1973, *Brick and block masonry*, BSI, 1973

BS CP 111: 1964, *Brickwork and blockwork walls*, BSI, 1964

BS CP 112 Part 2: 1971, *The structural use of timber*, BSI, 1971

Building Centre Trust and BRE, *Joints between concrete cladding panels. Building detail sheet pilot study*, The Building Centre Trust, 1970

BRE Digest 35: 1963, *Shrinkage of natural aggregates in concrete*, HMSO

BRE Digest 178: 1975, *Autoclaved aerated concrete*, HMSO, 1975

GENERAL REFERENCES TO ALL SECTIONS

BRE Digest 111: 1969, *Lightweight aggregate concretes 3: structural application*, HMSO, 1969

BRE Digest 237: 1980, *Materials for concrete*, HMSO, 1980

BRE Digest 59: 1970, *Protection against corrosion of reinforcing steel in concrete*, HMSO, 1965

BRE Digest 235: 1980, *Fixings for non-loadbearing precast concrete cladding panels*, HMSO, 1980

BRE Digest 157: 1973, *Calcium silicate (sandlime, flintlime) brickwork*, HMSO, 1973

BRE Digests 65, 66: 1966, *The selection of clay building bricks, 1 and 2*, HMSO

BRE Digest 58: 1965, *Mortars for jointing*, HMSO

BRE Digest 157: 1973, *Calcium silicate brickwork*, HMSO

BRE Digest 142: 1972, *Fill and hardcore*, HMSO

BRE Digest 123: 1970, *Lightweight aggregate concrete*, HMSO, 1970

The Concrete Society Technical Paper no 101, *The creep of structural concrete: a working party report*, The Concrete Society, 1973

The Concrete Society Technical Report no 14, *Guide to precast concrete cladding*, Report of a Joint Committee of the Concrete Society, Institution of Structural Engineers, Royal Institute of British Architects and British Precast Concrete Federation, The Concrete Society, 1977

C & CA, *Concrete Practice*, C & CA, 1975

Everett, A, *Mitchell's Materials*, Batsford, 1981

Greenwood, D A, *Mechanical improvement of soils below ground surface: Part 2*, ICE Ground Engineering Conference

Handisyde, C C and Haseltine, B A, *Bricks and Brickwork*, Brick Development Association, 1978

Harrison, T A, *Early-age temperature rises in concrete sections with reference to BS 5337:1976*, Interim technical note, C & CA, 1978

Hobbs, D W, *Shrinkage-induced Curvature of Reinforced Concrete Members*, C & CA, 1979

Morris, A E J, *Precast Concrete in Architecture*, George Godwin, 1978

Parrott, L J, *Simplified Methods of Predicting Deformation of Structural Concrete*, C & CA, 1979

Thomson, G H and Pearce, R W *Ground Treatment and Fill: Part 1*, ICE Ground Engineering Conference

Section 5:1, Foundations

Al-Hashirmj, K, *Foundation design: soil/structure interaction in shrinkable soils*, BDA Design note, Brick Development Association, 1977

Barnbrook, G, *House foundations for the builder and building designer*, C & CA, 1981

BRE Digests 63, 67: 1965/72, *Soils and foundations 1 to 3*, HMSO

BRE Digest 75: 1966, *Cracking*, HMSO

BRE Digest 142: 1972, *Fill and hardcord*, HMSO

BRE Digest 176: 1975, *Failure patterns and implications*, HMSO

BRE Digests 240–242: 1980, *Low-rise buildings on shrinkable clay soils parts 1 to 3*, HMSO

BS CP 2004: 1972

BS CP 110 Parts 1 to 3

BS CP 114: 1969

BS CP 101: 1972

CIRIA Technical Note 107, *Design for movement in buildings*, Alexander, S J and Lawson, R M, CIRIA, 1981

Greenwood, D A, *Mechanical improvement of soils below ground surface*, ICE Ground Engineering Symposium on Conference Paper, 1970

Thomson, G H and Pearce, R W, *Ground Treatment on Fills*, Parts 1 and 2, Cementation, Ground Engineering Ltd, Paper given at Concrete Symposium on ground treatment and Conference Paper Thurrock, 1976

Section 5:2, Basements and substructures and
Section 5:3, External walls

BS CP 102: 1973, *Protection of buildings against water from the ground*, BSI, 1973

BS CP 204: 1970 Part 2, *In-situ floor finishes*, BSI, 1970

CIRIA, *Guide to the Design of Waterproof Basements*, CIRIA, 1978

Deacon, R C, *Watertight concrete construction*, C & CA, 1973

Deacon, R C, *Concrete ground floors, their design, construction and finish*, C & CA, 1974

Perkins, P H, *Concrete ground floors for domestic buildings*, Advisory Note, C & CA, 1977

Perkins, P H, *Floor screeds: recommendations for cement-sand and lightweight screeds*, C & CA, 1968

Section 5:4, Structural frameworks

BS CP 110 Part 1: 1972, *The structural use of concrete*, BSI

BS CP 116: 1969, *Structural use of precast concrete*, BSI, 1969

BS CP 112: 1971 Part 2, *The structural use of timber*, BSI, 1971

Browne, R D, *Thermal movement in concrete*, current practice sheet, C & CA 1979

Hart, F, Henn, W and Soutag, H, *Multi-storey Buildings in Steel*, Granada publishing, 1978

Rodin, J, *The implications of movement in structural design*, proceedings at symposium of Concrete Society, The Concrete Society, 1969

Section 5:5, Brick and block masonry

BSI, See references to Section 3

Foster, D, *Brickwork: dimensional stability observations on movement*, Structural Clay Products, 1971

Tovey, A K, *Concrete masonry for the designer*, C & CA, 1981

GENERAL REFERENCES TO ALL SECTIONS

Section 5:6, Facings and claddings
Building Centre Trust and Building Research Station, *Joints between concrete cladding panels. Pilot study building detail sheet*, BRE and Building Centre, 1970

BRE Digest 196: 1976, *External rendered finishes*, HMSO, 1976 (see also references for sections 2, 3 and 4)

BRE Digest 199: 1977, *Predicted effects of inaccuracies on joints, fixings and bearings*, HMSO

BRE Digest 217: 1978, *Wall cladding defects and their diagnosis*, HMSO, 1978

BRE Digest 223: 1979, *Wall cladding designing to minimise defects due to inaccuracies and movements*, HMSO, 1979

BRE Digest 235: 1980, *Fixings for non-loadbearing precast concrete cladding parcels*, HMSO, 1980

BS 5606, *Predicted inaccuracies of cladding and structure*, BSI, 1982

The Concrete Society, *Paper on the provision of compression joints in the cladding of a reinforced concrete building*, The Concrete Society and C & CA, 1980

C & CA, *Practice note: External Rendering*, C & CA, 1970

Foster, D, *Further observations on the design of brickwork cladding to multi-storey RC frame structures*, Brick Development Association, 1981

CDA, *'Copper and copper alloy fixings for buildings'*, Technical note TN21 Sept 1975 reissued June 1976, Copper Development Association, 1976

Harris & Edgar, *'Joints between precast concrete cladding units'*, Paper given at C & CA by FBW/PHD of Trust Concrete Structures Ltd, 1975

Richardson, F A, *Design rules for cladding*: Paper in Stone Industries Journal, July/August 1969

Section 5:7, Roofs and roof finishes
BS, *Draft standard code of practice for flat roofs*, 77/11822 DC, BSI, 1977

Department of the Environment, *Construction*, PSA Quarterly Journal No 28, Property Services Agency Library, 1975–82

Department of the Environment, *Flat Roofs*, Technical Guide, Property Services Agency Library, 1981

MACEF, *Roofing Handbook*, Mastic Asphalt Council and Employers Handbook, 1980

Section 5:8, Internal suspended floors
See references to Sections 2–4 for British Standard Codes of Practice and for Building Research Establishment Digests 35 and 75.

Section 5:9, Internal finishes
Floor finishes

BRE Digest 47: 1964, *Granolithic concrete, concrete tiles and terrazzo flooring*, HMSO

BRE Digest 104: 1969, *Floor screeds*, HMSO

BRE Digest 75: 1966, *Cracking in buildings (floors)*, HMSO

BRE Digest 79: 1967, *Clay tile flooring*, HMSO

Wall finishes

BRE Digest 40: 1964, *Choosing specifications for plastering*, HMSO

BS 5492: 1977, *Code of Practice for internal plastering*, BSI, 1977

BS CP 202: 1972, *Tile flooring and slab flooring*, BSI

BS CP 203: 1969, *Sheet and tile flooring*, BSI

BS CP 204 Part 2: 1970, *In-situ floor finishes*, BSI

BS CP 211: 1966, *Internal Plastering*, BSI, 1956

Section 5:10, Components
BRE Digest 131, *Flexible drainage systems*, HMSO

Department of the Environment, *Construction quarterly*, Property Services Agency Library 1970–1982

Tremco Ltd, *Glazing systems: a guide to specification and use*, Tremco Ltd Construction & Products Division, 1979

Relevant National Standards
United States of America and Federal Republic of Germany

References to national standards which may be found useful in the following countries:

UNITED STATES OF AMERICA
American National Standards (ANS) 1430 Broadway, New York, NY10018

Concrete
ANSI/ACI 211.1–1977, *Selecting proportions for normal and heavyweight concrete practice* (Revision to ANSI A167.1–1974)
ANSI/ACI 211.2–1977, *Selecting proportions for structural lightweight concrete*, (Revision of ANSI A164.1–1969)
ANSI/ACI 531–1979, *Building code requirements for concrete masonry structures*
ANSI/ASTM C33–1979, *Concrete aggregates specification*
ANSI/ASTM–C5 12–76, *Method of test for creep of concrete*
ANSI/ASTM C426–70 (1976), *Test for drying shrinkage of concrete block*
ANSI/ACI 318–77, *Reinforced concrete building code requirements*
ANSI/ACI 223–77, *Recommended practice for the use of shrinkage-compensating concrete*
ANSI/ACI 322–71, *Building code requirements for structural plain concrete*
ANSI/ASTM D1850–67 (1972), *Specifications for concrete joint sealers*
ANSI/ASTM 01190–64, *Specifications for concrete joint sealer, hot-poured elastic type*
ANSI/ASTM D994–71 (1977), *Specification for preformed expansion joint filler for concrete*
see also ANSI/ASTM D1752–67 (1973), P1751–78 and D2628–69 (1976) for standards of preformed joint fillers

Timber
ASTM D52–62 (1976), *Wood paving blocks for exposed platforms, pavements, driveways and interior floors exposed to wet and dry conditions*
ANSI/ASTM D3503–76, *Test for swelling and recovery of compressed wood products due to moisture absorption*
D3043–76, *Testing plywood in flexure*

Steel
American Institute of Steel Construction: 1980. AISC Manual of Steel Construction

Building Code
International conference of building officials. Uniform Building Code (covers all aspects of concrete, wood, steel construction and seismic design)
ACI Manual of Concrete Practice 1979, Parts 1–3, The American Concrete Institute, Box 19150, Bedford Station, Detroit, Michigan 48219. Contains valuable guidance on joint placing, detailed design and selection of sealants

FEDERAL REPUBLIC OF GERMANY
DIN: Deutsches Institut für Normung e.V.
DIN 7340, 18201, 04.76, *Dimensional tolerances for building, definitions, principles, application testing*
DIN 7360, 18202 pt 1, 03.69, *Dimension tolerances in building construction, permissible allowances for the execution and the work wall and floor openings, recesses, store and landing heights*
DIN 7600, 4117, 11.60, *(Damp-proof, waterproof and weatherproof construction). Damp proofing of buildings against ground moisture; directions for construction*
DIN 7470, 18203 pt 1, 06.74, *Dimension tolerances in building construction, prefabricated parts of concrete and reinforced concrete*
DIN 7480, 52455 pt 2, 11.74, *Testing of materials for joint and glazing seals in building construction: adhesion and extension test*
52460, 08.79, *Sealing and glazing terms; calculation and constructional design*
DIN 4720, 1054, 11.76, *Subsoil: permissible loading of subsoil*
4019, 09.74, *Subsoil: settlement calculations for perpendicular central loading*

Note: The foregoing list is not a comprehensive list of all relevant German standards but a selection of those standards available in English translation.

Index

Basement construction 90
    details 94–96
    movement joints 93, 94, 95
    reinforcement 90
    retaining walls 92
    types 90–95
    wall junctions 96
    water retention 90, 94, 95–96
Basements 76, 90
Building damage, classification 67

Ceilings 160
    see also *Finishes*
Cladding/Facings 145–59
    design criteria 145–6
    failure limits 147
    fixings 153–7
    frame interaction 150–51
    joints 158
    movements 145
        assessment 147, 152
        effects 147
        failures 148, 150, 156
    Temperature fluctuations 146
Clay 70–71, 74
Clay fraction 70
    liquid limit, 70
    plasticity index 70
    types 71
Components 203–7
    draft proofing 205
    glazing 203–4, 206–7
    pipe junctions/joints 208–9
Concrete 23–34
    blockwork 34, 123
    creep 33, 34
    deviations 54–55, 159
    elastic deformation 31, 34
    early age movement 23, 28–30
    heat of hydration 23, 30
    plastic settlement 30
    shrinkage 32–34, 36–37, 156
    sulphate action 36
    thermal movement, 29, 30–32, 34–35, 165, 166
Control
    see *Control joints*
Control strategies 43

Control principles 43–44
Deflection
    beams 180
    floor slabs 177
    initial 182
    timber floors 187
Deformation 13–15, 118–19,
    adjoining elements 99, 118–19,
    bowing 16, 20
    deflection 67
    direction of 13
    distortion 106, 151
    elastic 15
    inherent 14
    linear 13–14
    panel walls 99
Deviations 51–57
    definition 53–55
    induced 53–55
    inherent 53–55
    permissible 54–55
External works 95
    paved areas 95, 98

Facings – see *Claddings*
Fill 71
    compaction 71
    movement 81
    settlement 71
    sulphates 81
    shale 85
    treatment 72
Fillers 58, 187
Finishes 190–203
    floor causes of failure 191
        ceramic tiled 192
        design criteria 190
        jointing details 194
        movement control 191–2
        poly vynil chloride 192
        screeds 191–3
        terrazzo 192
    roof 159–64
        bonding 161–4
        jointing details 170–75
        junctions 162, 173
        movements 161

        substrates 161–4
        trim 166
    wall 194–203
        design criteria 195–6
        junction details 198–201
        movement 196
        movement control 196
        tiling details 196–202
        types 196
Floors: suspended 168–83
    see also *Floor finishes*
    bearings 186
    classification 168
    concrete 174
    deflection 173, 177, 180–81, 182
    design criteria 168
    exposure 179
    failure limits 168, 177
    joints: details 189
    joints/location 188
    joints/major 185
    joint/requirements 184
    joint/sealants 187
    movement/causes 175–6
    lateral support 183
    movement/direction 178
    movement/effects 174–7
    movement/restraint 179–82
    serviceability limits 174
    timber 183
    types 168
Frameworks: structural 97–108
    bracing 101
    concrete 105
    cracking limits 102
    deformation 99, 150–51
    deformation limits 100
    design criteria 93, 100
    distortion 106
    joints: details 109, 110–11
    joints: location 107
    movements: causes 103
    movements: control 103–4
    movements: stiffening 104
    steel 105
    timber 105
Foundations 66–79
    building reputations 68

INDEX

design criteria 66–67
details 78
failure limits 67
movements: sources 67, 72, 74
narrow strip 76–77
piled 78
rafts 76
selections 74
settlement 74, 75
strip 74
types 74

Gaskets 204
Glass 203–4
　bowing 16
Glazing systems 204
　movement characteristics 27
　temperature changes 17
Ground floors 79–89
　crack control 87–89
　components 85
　damage classification 80–81
　failures 79
　joint details 88–89
　load bearing 83
　long strip 86–87
　slab failure 80
　slab settlement 79
　small slabs 83
　reinforcement 86
　visible damage 80

Humidity – see *Moisture*
Humidity, effect on concrete 34–36

Joinery 203
　door panels 203
　framing for wall linings 197
　junctions 198, 200
　moisture content 203
　moisture movement 203
Joints 46–65
　　see also *Sealants and fillers*
　clearance 52, 53
　components 61
　design procedure 46
　depth 53
　details 48–51, 95–96, 170–71,
　　172–3, 185, 188, 194–5, 197
　fire resistance 62
　functional requirements 64–65
　location 107, 116–17, 188
　methods 46–49
　profiles 46–49
　types 44–46
　width calculation 47, 51

Loading
　effects 15
　time of application 22
　see also *Elastic deformations*

Masonry – see *Walls*
　deflection 69
Modulus of Elasticity 15
　concrete types 31
　definition 15
　elasticity of concrete 33–34
Moisture 18, 25
　see also *Shrinkage and movements*
　humidity effects 39
　movements effects 18, 25
　　estimation 18, 25
　　in timber 36, 41
Movement
　accommodation 44, 105
　assessment 44, 112–13, 147
　causes/sources 12, 103, 160, 191
　characteristics of materials
　　24–27, 124–5
　classification 11–12
　differential/dissimilar 13, 120,
　　121
　direction 178
　early age 29–34
　estimation 15–22
　extrinsic 12
　free/unrestrained 11, 15, 162
　heat of hydration 23
　identification 122
　intrinsic 12
　linear 14
　moisture 18, 19
　in materials: brickwork 40
　　ceramics 41
　　concrete 23, 28, 37, 39
　　plastics 42
　　timber 36
　moisture 18, 190
　prevention 43
　sources – see *Causes*
　thermal 17

Restraint 14
　degree of 11, 16
　dissimilar 120, 121
　examples 179
　force of 11
Retaining walls 90–93
　basements 92
　failure: causes 90
　surface finish 90

types 90
Roofs 159–68
　see also *Finishes: roofs*
　critical points 162–3
　coverings 159
　design criteria 159
　failure limits 159, 160
　flat 159, 163, 164–8
　layers 159
　movements: assessments 164
　　causes 160–61
　　control 167
　　effects 160
　　prevention 166
　　repeated 162
　　restraint 165
　　time dependent 162
　joints: details 170–71, 172–3
　　formation 174–5
　　location of 167–8
　　systems 168

Sealants: 58–59, 60–61
　insertion 63
　performances 53
　properties 60
　selection 57–59
　types 187
Shrinkage 18
　see also *Moisture movements*
　drying in concrete 32–34
　in materials 26–27
　roof decks 164
　soils 70
Slip planes 133
Soils 67–73
　see also *Clay, fill*.
　characteristics 70
　movements 73, 77
　shrinkage 70
　swelling 70
　testing 70
　treatment 72
Sources: of
　movements/deformations 11
　classification of 11
　extrinsic 12
　intrinsic 12
Stress 11, 12, 18
　causes 11
　envelope: in soils 76
　estimation 18
　failing 19
　induced 19
　limits of 12
　magnitude 12

215

INDEX

types 11
Substructures 476
    see also *Basements*
Sulphates, in fill/hardcore 81
  in brickwork 112, 132

Temperature, base temperatures 17
  changes 17
  ranges for materials 17
  movements 17
Thermal movements, estimation 17
Tolerances 54–56
    see also *Deviations*

Tiling
  floor 192, 194
  wall 196, 197
Trees 73, 76, 77

Walls 108–45
  adjoining elements 118–9
  bearings 136
  cracks 108
  critical points 122–3
  damage 108
  design criteria 108
  failure limits 108

joints, control 125, 133, 135
  details 133, 138, 139
  formation 143
  location 113, 116
  movements 112, 113, 122, 124
  panel 137, 141
  parapets 112
  provision 133
  restraint 120–21
  sealants 143–4
  settlements 112
  types 125, 139, 140, 144
  weather resistance 143